Bioinoculants

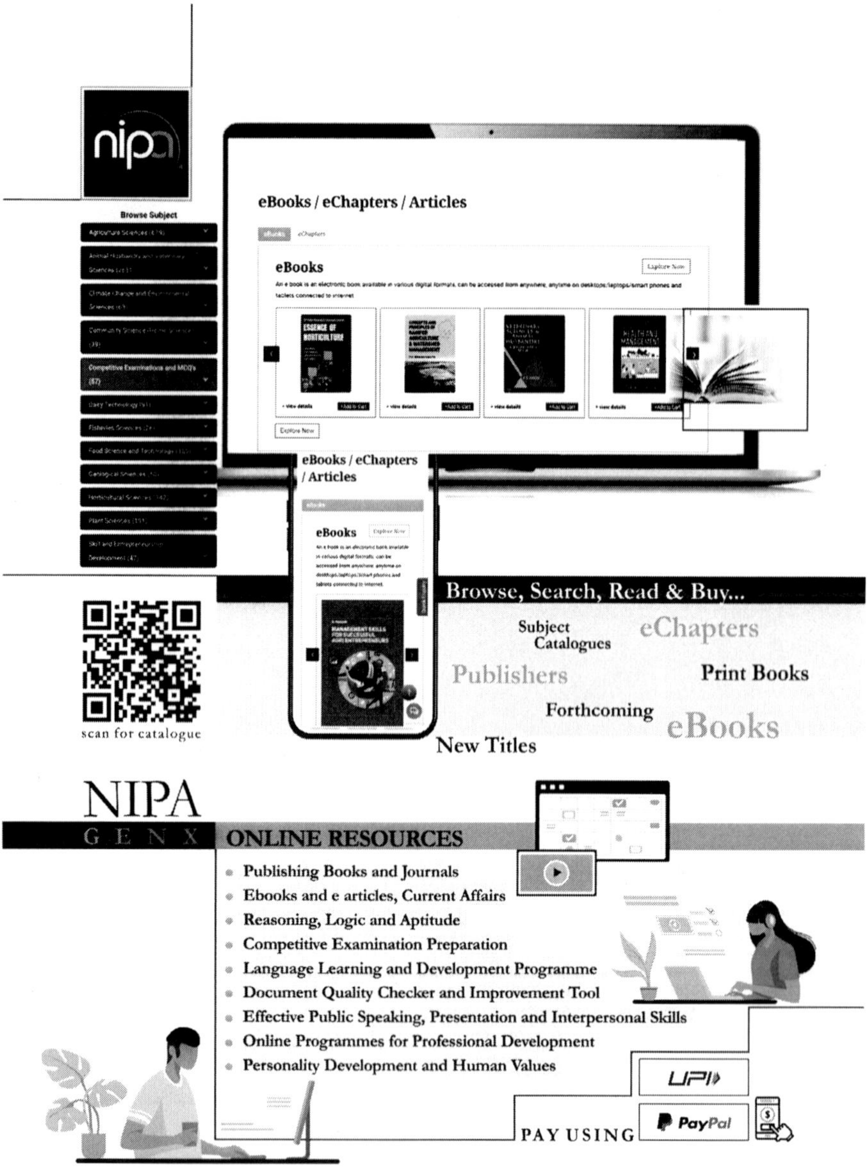
nipa
Browse Subject
eBooks / eChapters / Articles
eBooks
eBooks / eChapters / Articles
eBooks
scan for catalogue
Browse, Search, Read & Buy...
Subject Catalogues
eChapters
Publishers
Print Books
Forthcoming
eBooks
New Titles
NIPA
GENX
ONLINE RESOURCES
Publishing Books and Journals
Ebooks and e articles, Current Affairs
Reasoning, Logic and Aptitude
Competitive Examination Preparation
Language Learning and Development Programme
Document Quality Checker and Improvement Tool
Effective Public Speaking, Presentation and Interpersonal Skills
Online Programmes for Professional Development
Personality Development and Human Values
UPI
PayPal
PAY USING

Bioinoculants

A Step Towards Sustainable Agriculture

Authors

R.P. GUPTA

Coordinator Research
College of Basic Science and Humanities
Punjab Agricultural University, Ludhiana - 141 004
and Professor, Department of Microbiology
BIS Institute of Science of Technology
Gagra (Moga), Punjab

ANU KALIA

Scientist Department of Microbiology
Punjab Agricultural University, Ludhiana - 141 004

SHAMMI KAPOOR

Mycologist, Department of Microbiology
Punjab Agricultural University, Ludhiana - 141 004

New India Publishing Agency
Pitam Pura, New Delhi- 110 088

Published by
Sumit Pal Jain *for*
New India Publishing Agency
101, Vikas Surya Plaza, CU Block, L.S.C. Mkt.,
Pitam Pura, New Delhi- 110 088, (India)
Phone: 011-27341717, Fax: 011-27341616
E-mail: newindiapublishingagency@gmail.com
Web: www.bookfactoryindia.com

ISBN : 978-81-89422-21-9

Composed and Designed by NIPA

Preface

Bioinoculants represent the cheap alternatives to agro-chemicals that is the chemical fertilizers and pesticides. Keeping in view their role in sustainable agriculture, elaborate highlighting is required for projection and application of these inoculants in the coming days. The salient feature of this work is the contents or the topics covered which includes the biofertilizers as well as biopesticides, bioremediation and organic matter degradation. This book would provide the readers not just the comprehensive knowledge on the topic i.e. bioinoculants but is also equipped with information regarding the recent innovations and technological advancements in this sector of sustainable agricultural approach. We've tried to arrange the available up-to-date information on this topic so that it would appeal to the students of under graduate, graduate and post-graduate level as well as percolate to the minds of this class of students and the policy makers by virtue of its comprehensibility. Recent developments have been touched for all the subsects mentioned in the book under future perspective title. Moreover information is also provided in form of diagrams, pictures and if necessary as flow charts to make the subject clearer and easily assimilative.

We are thankful to Dr. B.S.Dhillon, Director of Research, PAU, Ludhiana and Dr.J.S. Dhaliwal, Director and Dr. M.S. Dhanju Principal of BIS Institute of Sciences and Technology, Gagra (Moga) for providing this opportunity and the necessary inspiration for the completion of the work.

R.P. Gupta
Anu Kalia
Shammi Kapoor

Ludhiana
June, 2007

Contents

Preface v

1 **Symbiotic Nitrogen Fixing Rhizobium** 1-35

2. **Actinomycetes Symbiosis With Higher Woody Plants** 37-60

3. **Cyanobacterial Symbiosis With Vascular Plants** 61-85

4. **Free Living, Loose Associative and Endophytic Diazotrophs** 87-114

5. **Phosphate Solubilizing Microorganisms** 115-138

6. **Vesicular Arbuscular Mycorrhizae** 139-181

7. **Plant Growth Promoting Rhizobacteria and Biocontrol Agents** 183-200

8. **Mass Production and Quality Control of Biofertilizers** 201-222

9. **Biopesticides** 223-256

10. **Organic Matter Decomposition** 257-271

11. **Bioremediation** 273-298

1

Symbiotic Nitrogen Fixing Rhizobium

- Introduction and History 2
- General Characters/Classification Criteria 3
- Rhizobial Taxonomy 4
- Morphology 4
- Cultural Characters 5
- Ecology 7
- Physiology and Molecular Biology 9
- Classes of Symbiotic Genes 10
- Symbiotic Megaplasmids (P*SYM*) 11
- Future Perspectives 27
- References 28
- Mind Bogglers 34

Introduction and History

Leguminosae/Fabaceae is a superfamily of dicotyledonous plants, which in species diversity is the third largest family of plants accounting for 5% of Earth's plant biodiversity and make up for more than at least 1% of the total Earth plant biomass. Among these, legume grains and pulses having more than 20 to 30% lysine protein content are cultivated in 145 million ha of land in developing countries. This family is also characterized by presence of specialized tumor like structure formation (nodules) on roots (occasionally on stem as in *Sesbania rostrata, Aeschynomene*) that is inhabited by microsymbiont gram-negative bacteria (*Rhizobium*), engaged in fixing atmospheric nitrogen or symbiotic biological nitrogen fixation.

Biological nitrogen fixation (BNF) is a unique prokaryotic phenomena estimated to be fixing 65% of the nitrogen (N) currently utilized in agriculture, and is expected to become increasingly important in future crop-productivity, particularly for sustainable systems (Thomas *et al.*, 1997). BNF provides more N to the agricultural ecosystems worldwide than the total amount of N fertilizer applied. BNF can be considered as "the" main entry of combined N in the N-cycle with estimates of 139-170x10^6 tNyr^{-1} equivalent to US $ 160-180 billion as compared to 80x10^6 t Nyr^{-1} for N fertilizers (Peoples and Craswell, 1992). However the estimated nitrogen fixed by BNF has been revised to 180 million tones per year nitrogen worldwide (Postgate, 1998).

Nitrogen fixation by legumes ranges between 25-75 kg of nitrogen per hectare per year in a natural ecosystem and several hundred pounds in a cropping system. Legumes in association with rhizobia may fix up to 250 kg of nitrogen per hectare and are not usually fertilized while perennial and forage legumes such as alfalfa, sweet clover, true clovers and vetches may fix 250-500 lbs of nitrogen per acre (Graham and Vance, 2000). Thus the symbiotic component contributes about 120 million metric tones year^{-1} to global nitrogen economy.

Some legumes are better at enriching soil nitrogen than others. **Common beans are poor fixers** (less than 50 lbs per acre) and fix less than their nitrogen needs. Other grain legumes such as **peanuts, cowpeas, soybeans,** and **faba beans** are good nitrogen fixers, and fix all of their nitrogen needs other than that absorbed from the soil.

The symbiotic associations have an important role in soil fertilization, improvement of growth and mineral nutrition. Hence, legume root nodules provide a built-in supply of nitrogen fertilizer. However there exists variation in the effectiveness of different rhizobial strains in legume rhizobial association.

History of major discoveries in rhizobial symbiosis

- **Leonhard Fuchsius**, a German physician and botanist in 1542 published drawings of nodulated legumes
- **Malpighi**, an Italian worker (1675), observed nodules on the roots of beans (*Phaseolus vulgaris* and *Vicia faba*).
- **Woronin**, a Russian botanist (1866) noted that the nodules of *Alnus glutinosa* (Betulaceae) and *Lupinus mutabilis* (Leguminosae) were filled with minute bodies resembling bacteria.
- **Frank**, a German researcher (1879) found nodules on the roots of all healthy legumes in a study and demonstrated that incinerating soil prevented the nodulation of *Pisum sativum*.
- **Hellriegel** (1889) and **Hellriegel and Wilfarth** (1890) both German researchers showed that nodule formation results from external infection of *Lupinus* spp., *P. vulgaris, P. sativum, Ornithopus sativa, Trifolium* spp., and *Vicia sativa*.
- **Bejerinck**, a Netherlands researcher (1888, 1890), first proved that bacteria provoke nodule formation
- **Prazmowski**, a Polish researcher (1889, 1890) showed bacterial penetration of root hairs of legumes by infection thread formation during inoculation of *P. sativum* with pure cultures

General characters / Classification criteria

Rhizobial systematics involves several criteria i.e. a variety of features for description of a new bacterial genus and species as growth rate on laboratory media, carbohydrate utilization patterns, metabolic capacities, antibiotic sensitivity, total protein fingerprinting, multilocus enzyme electrophoresis, fatty acid analysis, DNA G+C content (Graham *et al*, 1991) though host range is still an important criterion.

Rhizobial taxonomy

Classically rhizobia are defined as the symbiotic bacteria capable of forming tumor-like outgrowths from pericycle followed by invasion by infection thread formation to form nodules that exhibit symbiotic nitrogen fixation. Fred *et al* (1932) recognized six species of *Rhizobium* on basis of host infectivity and certain biochemical tests into various cross-inoculation groups. *Rhizobium* genus contains members that are Gram negative motile rods, aerobic chemoheterotrophs having 59-65.5% G+C (Table 1).

Table 1. Classical systematics of rhizobia

Species	Host Nodulated
Group I: Peritrichous flagella (2-6); GC content of DNA (59.1-63.1%) Fast grower	
Rhizobium trifolii	*Trifolium*
R. leguminosarum	*Pisum, Lens, Vicia, Lathyrus*
R. phaseoli	*Phaseolus vulgaris*
R. meliloti	*Melilotus, Medicago*
Group II : Polar or subpolar flagellum; GC content (61.6-65.5 %) Slow grower	
R. japnicum	*Glycine max*
R. lupini	*Lupinus, Ornithopus*
Rhizobium sp.	*Vigna, Arachis*

Morphology

A typical rhizobial cell is small to medium sized (0.5 to 0.9x1.2 to 3.0 μm) gram negative rod or bacilli, motile when young and cells exhibit characteristic presence of copious amounts of β-hydroxybutyrate granules forming 40-50% of the cell dry weight easily observed by using stains for metachromatic granules. Most strains produce sticky gum like substance of varying composition.

CULTURAL CHARACTERS

Rhizobia show a typical translucent, viscid, slimy growth on YEMA media with individual colonies having dome shaped, elevated feature with entire margins. Rhizobia grow better on glutamate supplemented media than nitrate or ammonium ion supplemented media. Rhizobial growth is also affected by presence of other inorganic cations as calcium deficiency in presence of magnesium reduces *R.trifolii* growth because calcium deficient cells become swollen and vacuolated. Low concentration of vanadium, cobalt and iron (at 10ppm concentration) stimulates growth of some rhizobial species. On basis of appearance of growth on YEMA petri plates the rhizobia are grouped as fast growers and slow growers with the latter group to be more fastidious in carbon source as well as vitamin requirements. Slow growers can utilize mannitol, arabinose, xylose, galactose, fructose, sodium citrate and rarely dextran. Vitamin requirements are also substantially diverse and include thiamin, biotin and calcium pentothenate while no response is observed for nicotinic acid, pyridoxin, folic acid, p-amino benzoic acid, inositol, Vitamin B_{12} and riboflavin.

Molecular basis for reorganization of group rhizobia

Polyphasic taxonomical approach has been instrumental in classifying rhizobia in seven genera and 36 rhizobial species (Sahgal and Johri, 2003, Wang and Martinez-Romero, 2000; Sy *et al*, 2001). Early nineties came up with the sequence comparison of 16S rRNA genes and genetic fingerprinting methods based on PCR that culminated to the characterization of rhizobia. The approach provided new insight for *Rhizobium* classification as it used the phylogenetic relationship among bacteria. The topology of the phylogenetic trees obtained from aligned sequences of full sequence of 16S rRNA (ssu rRNA) and ARDRA patterns were found to be most reliable to ascertain the phylogeny of the rhizobial isolates. Rhizobia have been divided into three genera viz. *Azorhizobium, Bradyrhizobium* and *Rhizobium* (Willems and Collins, 1993). Subsequently the genus has been further classified into two new genera, *Mesorhizobium* and *Sinorhizobium* (de Lajudie *et al*, 1994; Jarvis *et al*, 1992) while *Rhizobium leguminosarum* remained type species and defined the genus (Young, 1996). However *R. phaseoli* and *R. trifolii* differed in host range from the above thus currently, *R. leguminosarum* has 3 biovars viz. *viciae, trifolii* and *phaseoli*. A new genus *Allorhizobium* has been proposed for nitrogen fixing species of *A.undicola* accommodating isolates which effectively nodulate *Neptunia natans* (Table 2).

Table 2. The reorganization of rhizobial members (rhizobial taxonomy)

Genus	Species	Host (s)
Allorhizobium	*A.undicola*	*Neptunia natans*
Azorhizobium	*A. caulinodans*	*Sesbania rostrata*
Bradyrhizobium	*B.elkanii*	*Glycine max*
	B.japonicum	*Glycine max*
	B.liaoningenese	*Glycine max*
Mesorhizobium	*M. amorphae*	*Amorpha fruticosa*
	M. chacoense	*Prosopis alba*
	M. ciceri	*Cicer arietinum*
	M. huakuii	*Astragalus*
	M. loti	*Loti*
	M. mediterraneum	*Cicer arietinum*
	M. plurifarium	*Acacia, Leucaena*
	M. tianshanense	*Glycyrrhiza, Sophora, Glycine*
Methylobacterium	*M.nodulans*	*Crotolaria pedocarpa*
Rhizobium	*R.etli*	*Phaseolus vulgaris*
	R.galegae	*Galgae*
	R.gallicum	*P.vulgaris*
	R.giardinii	*P.vulgaris*
	R.hainanense	*Centrosema,Desmodium,Tephrosa*
	R.huautlense	*Sesbania herbacea*
	R.leguminosarum	*Trifolium, Vicia*

Contd...

Table 2. Contd...

	R.mongolense	*Medicago ruthenica*
	R.phaseoli	*P.vulgaris*
	R.sullae	*Hedysarum hedysari*
	R.tropici	*Leucaena, P.vulgaris*
	R.trifolii	*Trifolium*
	R.yanglingenese	*Amphicarpaea,Trisperma, Corollina varia,Gueldenstaedtia multiflora*
Sinorhizobium	*S.arboris*	*Acacia sensgal,Prosopis chilensis*
	S.fredii	*Glycine max*
	S.kostiense	*Acacia senegal, P.chilensis*
	S.medicae	*Medicago spp.*
	S.meliloti	*Medicago sativa*
	S.saheli	*Sesbania*
	S.terangae	*Acacia, Sesbania*
	S.xinjiangense	*Glycine max*

ECOLOGY

Rhizobia include the non-sporeformer gram negative rods that exhibit a strange feature of survival in soil for long periods ranging from 19 to 45 years though the number declines in absence of the host plant roots.

Biotic factors: The rhizobia exhibit variation in their number and other characteristics as a result of interaction with the biotic factors i.e. the resident indigenous soil microbial population; which includes synergistic/antagnostic interactions with many fungi and some bacilli like *Pseudomonas, Bacillus sp.*. Culture filtrates of fungi isolated from soil often exhibit rhizobial growth. Among these antibiosis interactions the slow growing rhizobia show more susceptibility than the fast grower rhizobia. However the strains that produce rhizobiotoxins have an extra edge for better survival and winning the space and food in competition with other soil microbes near/in rhizosphere of plant roots. Production of antibiotics such as aureomycin, terramycin and ledermycin affects the number of rhizobial colonies/cells. These also get easily eliminated by bacteriophages i.e. the rhizophages when present in soil solutions.

Rhizobia cherish a wide host range for example *Rhizobium* sp. strain NGR234 has a host range of more than 112 genera of legumes (Viprey *et al,* 2000). These symbiotic relationships occur when rhizobia penetrate their hosts with centripetally-developing infection threads.

Abiotic factors: Rhizobia exhibit a range of temperature sensitivity and occur in diverse regions including very low temperature hilly areas as well as moderate to high temperature plains/terrains. *R. lupini* and *R. japonicum* are comparatively resistant to high soil temperatures while the rest of genera show sensitivity to increase in temperature above 30°C. Tropical rhizobia that nodulate *Acacia, Lotus* and *Psorales* have adapted to their harsh temperature regimes. The alfalfa group of rhizobia are relatively more tolerant to high temperature than those that nodulate pea, clover and bean.

Soil pH status also affects rhizobial population in rhizosphere particularly the soil acidity that minimizes the population while alkaline pH stimulates the growth of rhizobia. This feature is observed in the YEMA media required for rhizobial cultivation that has a slight alkaline pH. The buffering action of calcium carbonate or calcium hydroxide neutralizes the acidic ions in acidic soils and help in stimulating growth of rhizobia in acid soils. Among rhizobial genera *R.meliloti* is found to be very acid sensitive while *R.japonicum* is able to tolerate pH drop as low as 3.5.

Xenobiotics and pesticides particularly fungicides and weedicides in soil have varied effects on rhizobia. The above two categories of pesticides are observed to be more toxic to rhizobia and thus decrease the soil inoculum. But overall the susceptibility towards the affect of these chemicals differs a lot among different abiotic and biotic conditions. *R.lupini* and *R.trifolii* are most sensitive to 2,4-dichlorophenoxy butyric acid and 4-chloro, 2-methyl phenoxyacetic acid while *R.meliloti* is comparably more resistant to their toxic action.

Techniques for ecological studies: Ecology of rhizobial cells as bacterial symbiont has lagged behind, largely due to a lack of practical techniques that can be used to monitor and assess the performance of these bacteria in the field. Most techniques developed in the last few decades involved 'marking' individual strains so as to follow the fate of inoculated strains in field environment. The marking strategies provide knowledge of success or failure of specific strains in a range of environments, however these provide meager information on nature of the pre-existing rhizobial populations or interaction between marked strains and the background population at these sites. The advent of molecular techniques such as PCR based T-RFLP, DGGE, AP-PCR, 16s rDNA techniques have revolutionized the study of *Rhizobium* ecology by allowing us to follow the flux of a variety of ecotypes within a particular site and to examine how introduced rhizobia interact with a genetically diverse background. In addition, molecular techniques have increased our understanding of the genetic evolution of the introduced strains in field sites and the response of individual strains and populations of root-nodule bacteria to changes in the environment.

PHYSIOLOGY AND MOLECULAR BIOLOGY

Nitrogen metabolism in rhizobia involves an 'inducible' nitrogenase enzyme and the rest of electron transporter proteins; that form a complex to carryout the reduction reaction in nodules. *Rhizobium* needs several gene products to perform symbiotic nitrogen fixation such as:

- '*nod* genes' which synthesize nod factor that initiate nodule development
- '*nif* genes' synthesize nitrogen-fixing apparatus
- '*fix* genes' synthesize products required in respiration and metabolism of bacteroid cells

Both '*nif*' and '*fix*' genes either occur on chromosome as in *B.japonicum* or on large symplasmids (Sym A and Sym B) in *R. meliloti*. The '*nif*' genes are structurally homologous to 20 *K.*

pneumoniae nif genes while '*fix*' genes which are involved in symbiotic association are not present in *K. pneumoniae.* The '*nod*' genes transcribe in presence of a transactivator *nod D* gene product. i.e. NodD protein which belong to a family of LysR-like transcriptional regulators and binds to highly conserved 47-bp DNA motifs (*nod* boxes) found in the promoter regions of many nodulation loci (Fisher *et al,* 1988). Apart from *nif* and *fix* genes the fine tuning and performance of the nitrogenase enzyme complex functioning could be enhanced by presence of *hup* genes or hydrogenase genes.

Classes of Symbiotic Genes

Class I - Genes for Nodule Initiation

- Nodulation genes - *(nod)*

Class II - Genes for tissue and cell invasion

- Acidic extracellular polysaccharide (*eps)* genes;
- Lipopolysaccharide (outer membrane) *lps* genes;
- β-1,2 cyclic glucans (periplasmic) *ndv* genes;

Class III - Genes for Nodule Function

- *nif* genes - nitrogenase determinants of *Klebsiella*
- *fix* genes – involved in biochemical reactions of bacteriods
- *dct* genes - dicarboxylate transport
- *cyt* genes - cytochrome determinants

Genomic organisation of nif and fix gene clusters

Rhizobial *nif* genes are structurally homologous to 20 *Klebsiella pneumoniae nif* genes. At least 9 different rhizobial nif genes have been identified in *R. meliloti, B. japonicum* and *A.caulinodans.* The *nif* and *fix* genes of above three rhizobia are organized in distinct clusters whose structure and genomic location are species specific. *R.meliloti* carries two megaplasmids of about 1400 kb (pSym-a) and 1700 kb (pSym-b). Both cluster I (*nif HDKE,* fix *ABCX, nifA, nif B, fdx*) and cluster II (*fix LJ, fix K, fix NOQP, fix GHIS*) are located on pSym-a. The cluster-II genes map at about 220 Kb downstream of *nif HDKE* operon and transcribed in opposite direction. A cluster of nod genes (*nod ABC*) is located in the 30 kb region between *nif E* and *nif N* (Fig 1). *B.japonicum* and *A.caulinodans* do not have plasmids and *nod, nif* and *fix* genes in these microbes are located on chromosome.

SYMBIOTIC MEGAPLASMIDS (P*SYM*)

- P*sym* of *R. meliloti* is 1200-1500kB, > 10% of genome
- In *Azorhizobium* and *Bradyrhizobium* spp, *sym* genes are chromosomal.
- In *Mesorhizobium loti* a "symbiotic island" is inserted into a tRNA gene.
- *nif* genes are also closely linked to *nod* genes on Sym plasmid.
- genes as *eps, lps, dct* are often found on plasmids other than P*sym*

Host plant root

↓

Secretion of flavonoids or signal compounds in the rhizosphere

↓

Induction of the nod operon

↓

Transcription of the *nod D* gene

↓

Switch on of the *nod ABC* genes and other nod genes

↓

Formation of the chitooligosaccharides or Nod factors

↓

Induction of the early nodulin genes in plant host

↓

Induction of root hair curling and cortical cell division in target root cells

Initial events in the preliminary signal interaction reaction in host plant and rhizobia

Three of these *nif* genes (*nif KDH*) codes for nitrogenase enzyme containing two characteristic components.

1. a molybdenum-iron-containing protein which reduces substrate coded by *nif D* and *nif K* genes, specifying α and β subunits of $\alpha_2\beta_2$ FeMo protein (component I or dinitrogenase Mr~ 220,000).
2. An iron-containing protein, which transfers electrons to the molybdenum-iron-containing protein, encoded by *nif H* (Homodimeric Fe protein, component II or dinitrogensae reductase Mr~ 60,000).

Fix genes are organized in a single operon in all except *B. japonicum* in which *fix A* and *fix BCX* form distinct transcriptional units present in cluster II and I. The *fix ABCX* gene products might be involved in electron transport to nitrogenase. Fix X contains 5 conserved cysteines of which 3 are arranged in as Cys-X-X-Cys-X-X-Cys cluster typical of feredoxins.

The *fix NOQP* genes when deleted in *B. japonicum* and *R. meliloti* strains show defective symbiotic nitrogen fixation. In addition *B. japonicum* mutants are affected in bacteroid development and exhibit a decreased whole-cell oxidase activity when grown microaerobically or anaerobically. The predicted a sequence of Fix NOQP proteins implies that they constitute a membrane-bound cytochrome-c containing heme/copper cytochrome oxidase which plays an instrumental role in bacteroid respiration under low oxygen conditions prevailing in root nodules.

The *fix GHIS* gene products are predicted to be transmembrane protein as Tn5 insertions in *fix G* or *fix I* genes result in a reduced symbiotic nitrogen fixation activity and marginal effect on nitrogenase activity under free-living conditions. This cluster of *fix* genes are likely to be involved in a redox process that contain 2 cysteine cluster typical of Fe-S centers present in bacterial ferredoxins. Fix I is homologous to catalytic subunit of bacterial and eukaryotic ATPases involved in cation pumping.

The expression of nitrogen fixation genes is controlled by cascades of hierarchically organized regulatory genes. Their concerted action enables bacteria to sense optimal environmental conditions required for nitrogen fixation and to transmit this information to level of genes expression. Down regulation of activity of nitrogenase at post-transnational level in response to presence of combined nitrogen. This control is based on the reversible ADP-ribosylation of nitrogenase reductase and operates in *A. caulinodans*. Environmental nitrogen and oxygen condition are critical signals for regulation of *nif* gene expression in *K. pneumoniae*. This dual physiological control is mediated via general nitrogen regulatory system (*ntr*) and *nif*-specific regulatory genes, *nif LA*. By contrast, expression of *nif* genes in symbiotic diazotrophs is regulated predominantly in response to cellular oxygen conditions. A largely identical regulatory system (*fix L, J, K, nif A, RpoN*) is used however theses are integrated into different species-specific networks. Two regulatory cascades control different groups of target genes. The cascade including Nif A controls expression of nitrogenase structural genes while genes encoding accessory functions is controlled by RpoN-dependent –24/12 promoters.

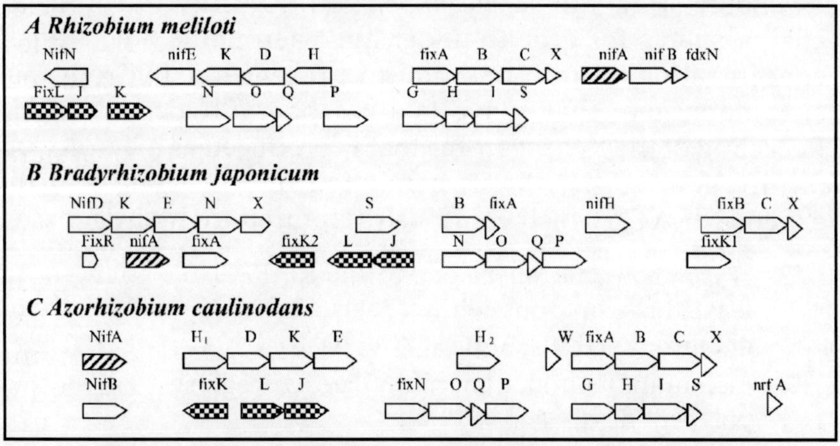

Organization of *nif* and *fix* genes clusters (Fisher, 1994)

Hydrogenase and hydrogen metabolism

Uptake hydrogenases allow rhizobia to recycle the hydrogen generated in the nitrogen fixation process within the legume nodule. Hydrogenase (*hup*) systems in *Bradyrhizobium japonicum* and *Rhizobium leguminosarum* bv. *viciae* show highly conserved sequence and gene organization, but important differences exist in regulation and in the presence of specific genes.

Many organisms have more than one hydrogenase, and many hydrogenase consist of more than a single polypeptide chain. Some hydrogenase are active only in the uptake of hydrogen from the atmosphere, whereas others, depending on the conditions, can also synthesize hydrogen.

$$N_2 + 8H^+ + 8e^- + 16\ MgATP = 2NH3 + H_2^- + 16MgADP + 16Pi$$

During nitrogen fixation, H^+ is reduced to H_2^- gas by nitrogenase and energy in the form of ATP is wasted on the production of hydrogen that is lost to the atmosphere. The efficiency of nitrogen fixation is lowered because only 40–60% of the electron flux is transferred to N_2. The challenge therefore is to recycle H_2 gas to H^+ to increase the efficiency of the process.

Hydrogen Metabolism

Some strains of *R. japonicum* could use hydrogen as an energy source for growth under microaerophilic (low oxygen concentration) conditions due to presence of hydrogenase enzyme system that converts atmospheric hydrogen to H^+. Recycling of the hydrogen gas that is produced as a by-product of nitrogen fixation. H_2 is generated by nitrogenase at the expense of ATP which could be recaptured by hydrogenase.

- 17 common genes (*hupSLCDFGHIJKhypABFCDEX*) arranged in at least three operons with conserved gene composition and organization (Ruiz-Argüeso *et al,* 2000)
- membrane-bound, dimeric [NiFe] hydrogenase having two polypeptides (35 and 65 kDa).
- *hupS* and *hupL* gene products are Hydrogenase structural subunits

- ❒ remaining *hup* and *hyp* gene products are involved in the recruitment and incorporation of nickel and other metallic groups into the hydrogenase active site

---

Legume nodulation by rhizobia

Symbiosis with legumes arose independently in two or more lineages of alpha proteobacteria. The deepest divisions occur between the *Bradyrhizobium* group (closely related to *Blastobacter, Rhodopseudomonas,* etc), the *Mesorhizobium-Rhizobium* group (closely related to *Agrobacterium*) (Wang and Martinez-Romero, 2000), and the newly discovered *Methylobacterium nodulans* (Sy *et al*, 2001).

The rhizobia are unique plant symbionts as they have evolved specialized infection through unicellular extensions of the modified epiblema at the root apex i.e. root hairs. These small single cell extensions get extensively colonized by soil-borne microorganisms due to their ability to exude several organic and signal compounds that help in recognition of the two symbiotic partners.

- ❒ Mimosoid legumes harbor broad host range rhizobia and are capable of interacting with diverse bacteria
- ❒ Members of tribe Phaseoleae of family fabaceae that includes annual and climbing legumes, have most broad-spectrum hosts

Development of nodules on legume roots involves an array of intricately fine choreographed "handshaking" or "lock and key" steps of signaling cascade reactions however the landmark signals could be summarized as three different sets of symbiotic signals that are prerequisite for identification between two symbionts during exchange process and help in determining specificity:-

1. First set of symbiotic signals include secretion of **Flavonoids/**phenolic compounds from plant roots in the rhizosphere where these compounds interact with the rhizobial Nod D proteins and function as environmental sensors and activators of transcription as only the correct

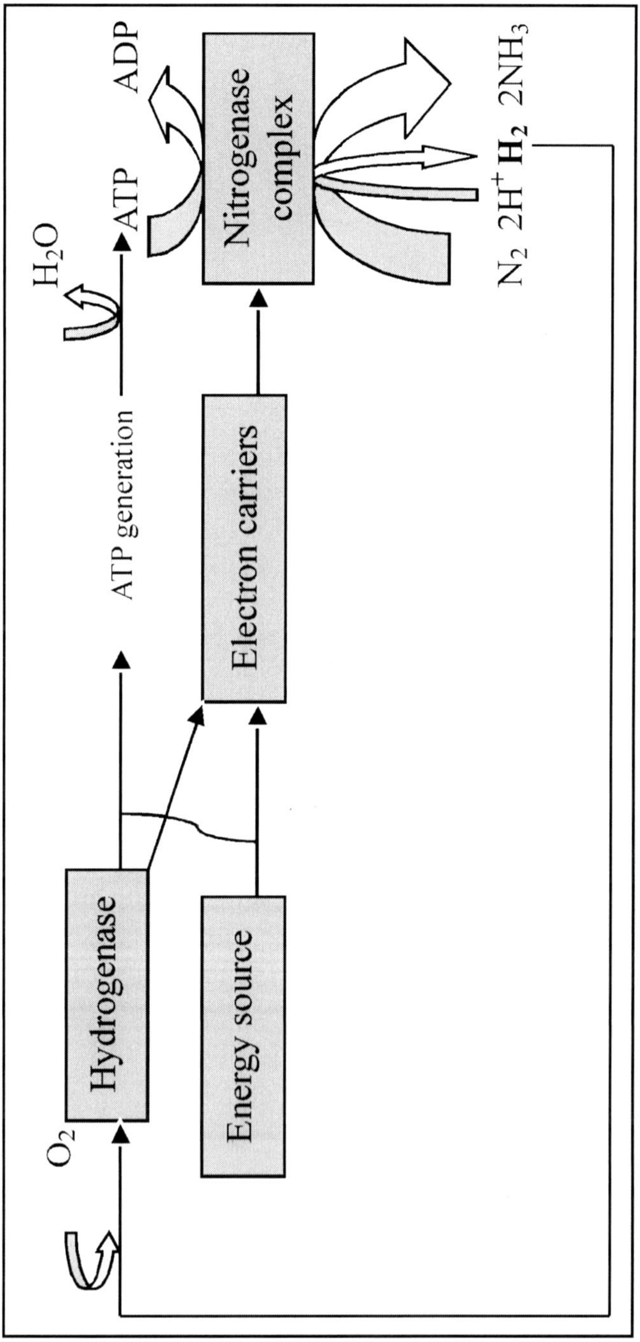

Recruiting of electrons and protons for ATP generation by hydrogenase enzyme complex

flavonoid(s) induce symbiotic gene expression in a particular rhizobial strain. Legume roots secrete a variety of (iso)flavonoids (phenolics required as secondary products in defense related reactions) which induce symbiotic genes in homologous bacteria (Perret *et al*, 2000). These phenolics are perceived as aglycones (2-phenyl-1,4-benzopyrone derivatives) that are among the 19% of photosynthate released as organic material into the rhizosphere by legume roots along with carbohydrates, organic acids, vitamins and amino acids. The flavanoids produced by particular host plant form the first step of bacterial recognition of the plant choice.

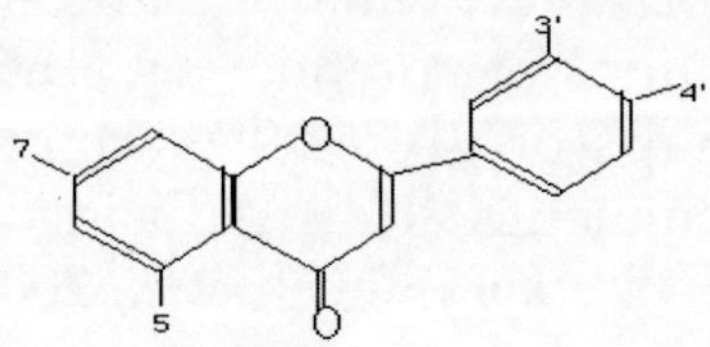

Flavone

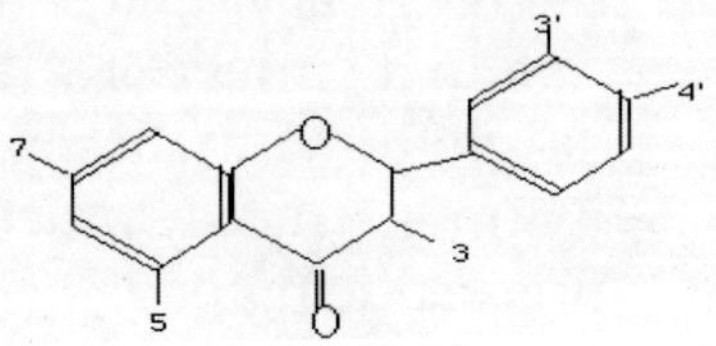

Flavone

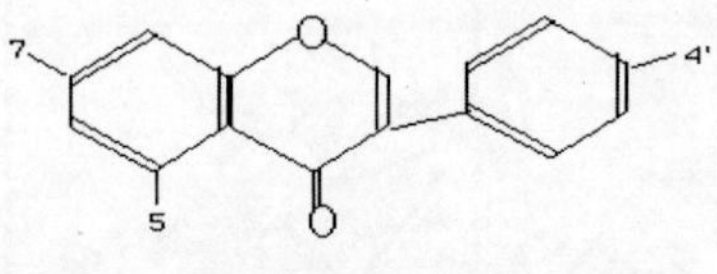

Isoflavone

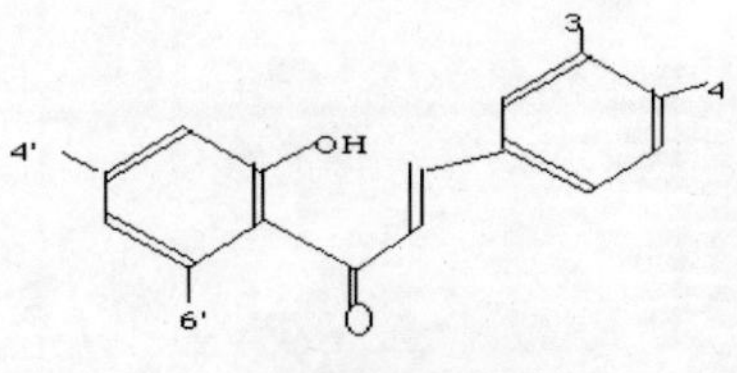

Chalcone

Common structures of various phenolics that act as signal compounds

2. The second set includes synthesis of **substituted lipo-chitooligosaccharidic Nod factors** by transcription and translation of *nod* box or *nod* genes by binding of the NodD-flavonoid complex. The flavanoids binding to the *nodD* gene product (Lys-R type regulator) results in conformational changes in NodD protein that favours its binding to the promoter region of the *nod* genes (enhanced transactivator properties). The nod proteins are various enzymes involved in the synthesis of substituted lipo-

chitooligosaccharidic **Nod factors** to provoke deformation of root hairs and allow rhizobia to enter the root through infection thread formation. The *nod* gene expression is observed to be probably related to sequence variations in individual *nod* boxes and this attribute also help in fine tuning the nod protein function/amount. Bacterial nod factor backbone consists of chitin subunits (four to five N-acetyl glucosamine units) linked by β-1-4 linkage and a fatty acid.

However variations have been reported in number of sugar backbone which may be longer than five subunits or the type of fatty acid moiety attached to the sugar, degree of acyl unit saturation and chemical modification due to a variety of substitutions/decorations by acetate, sulphate or carbomyl groups or glycosyl (arabinose, mannose, fucosyl, methyl) moieties at the reducing and non-reducing ends of the backbone (Hirsch *et al,* 2001).

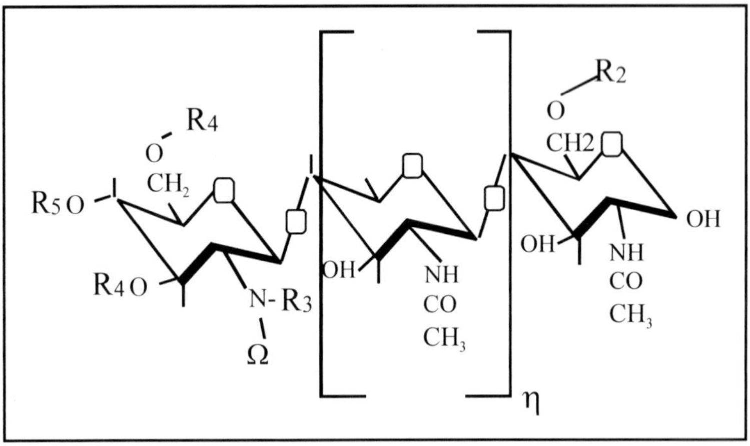

Structure of lipo-chitooligosaccharide made of N-acetyl glucosamine pentamer decorated by substitutions at various reducing or non-reducing ends

3. The third set of symbiotic signals include the secretion of **rhizobial products** which may have a major role in infection thread development (continuous) and include

extracellular polysaccharides (EPS)/related compounds and proteins exported by the type three secretion system (TTSS).

- ❐ Nod factor product (lipochitooligosacharide) signal is active at femtomolar concentrations
- ❐ Nod factor product has a structure correlated with host specificity of symbiosis suggesting the involvement of a cognate perception system in plant host
- ❐ 19 *nod* boxes on symbiotic plasmid of the broad-host-range *Rhizobium* sp. strain NGR234

The process of root hair curling is initiated to further result in infection thread formation and its ramification in the cortical region of the host legume root if there occurs exact synchronization of respective signal events between both symbionts correctly. Rhizobia exhibit partner choice owing to secretion of variety of nod factors at their end and polysaccharide composition of the rhizobial cell walls as well as the variety of root exudates (phenolics) secreted and discrimination of nod factors by the host legume. On reaching a stringent compatibility the bacteria multiply in the growing/branching infection thread and move into the root cortex resulting in development of a nodule structure. On reaching the inner tissues of the nodule the rhizobia undergoe a variety of morphological, physiological and biochemical changes to perform the function of nitrogen fixing minifactory. Rhizobial cells become intracellular are get surrounded by peribacteroid membrane, differentiate into bacteroids, change shape and derepress nitrogenase and the auxiliary enzymes required for reduction of the atmospheric dinitrogen. However, bacterial release from the infection thread can be blocked in incompatible interactions, that advocate the occurrence/requirement of specific signals to plant at this stage of symbiosis too (Perret *et al*, 2000).

The rhizobia fulfill their energy requirement from the host plant in lieu of the nitrogen provision to the plant in form of dicarboxylic acids (malate and succinate). Moreover the host plant also ensures the development and maintenance of the microaerobic conditions for better production and functioning of the O_2 sensitive nitrogenase enzyme complex by packing of nodule tissue with O_2 scavenging leghaemoglobin protein. About 16 to 42 ATP are utilized to reduce a single molecule of nitrogen to ammonia which is assimilated to surrounding host tissues possibly *via* an alanine shuttle (Waters and Emerich, 2000). The requirement of high amount of ATP energy during bacteriod development and respiration calls for physiological enhancement in production of cytochrome heme (Fe-S) proteins. Moreover an altered electron transport pathway is expressed including a cytochrome oxidase that can function in the low-O_2 environment of a nodule.

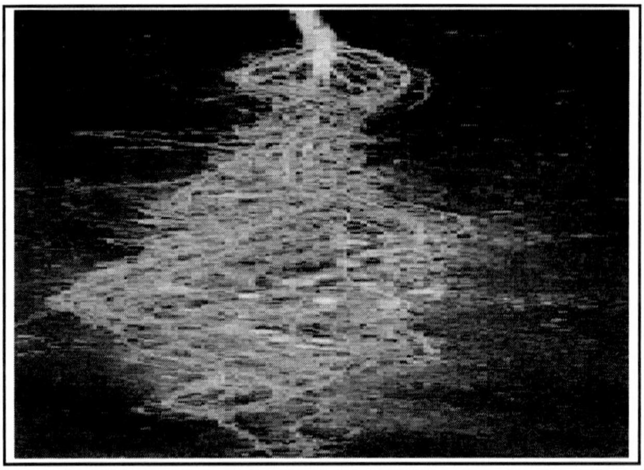

Formation of nodule on tap and lateral roots of soybean plant
(For a coloured version of this figure, see plate section, page 299)

Summarizing the Characteristic Features of Legume-Rhizobium symbiosis

- Secertion of flavanoids by host plant roots
- selective chemotaxis of rhizobia towards signals
- morphological specialisations on host root hair and rhizobial cell wall surfaces

- ❒ close cell-cell contact at interface (enlarged surface area; thin cell walls)
- ❒ polar adsorption of rhizobia on root hair surface
- ❒ initiation of root hair curling
- ❒ invasion of rhizobia and depolymerisation of cell wall matrix
- ❒ deposition of new wall material vesicles around invading bacteria
- ❒ infection thread formation
- ❒ suppression of host defence responses (as synthesis of phytoalexins, PR proteins)
- ❒ cortical cell division
- ❒ formation of nodule primordia
- ❒ movement and branching of infection thread
- ❒ release of rhizobia in cortical cells
- ❒ morphological and physiological alternations in rhizobial cell
- ❒ expression of novel proteins as nitrogenase protein complex (product of ~21 genes) in microsymbiont and leghaemoglobin in host cells
- ❒ metabolic integration
  - bacteria gain carbon supply from photosynthate for respiration;
  - plant provides microaerobic tissue (O_2-protection of nitrogenase);
  - Nitrogen-fixing bacteroids excrete NH_3
  - Host cells assimilate NH_3 through GS/GOGAT pathway (*glutamine synthase-glutamine-oxoglutarate amino transferase*)
  - Translocation of fixed-nitrogen as amides or ureides.

Genetic manipulation of legume or rhizobial host for understanding of plant-microbe interaction

a) *Legume genetic manipulation*

The genetics of legume-*Rhizobium* symbiosis has gained impetus after reports of isolation and molecular characterization of symbiotic receptor protein family in legumes (Endre *et al,* 2002; Starcke *et al,* 2002). These reports emphasize the occurrence of peculiar and universal symbiotic receptor protein family members that contain three distinct domains *viz.*

a) **an extracellular leucine rich repeat** (LRR) resemble Toll like receptor proteins in animals that work as antigen receptor proteins involved in innate immune response

b) **a hydrophobic transmembrane domain** joining LRR to kinase domain

c) **an intracellular serine or threonine kinase domain** that triggers signaling cascade inside the cell

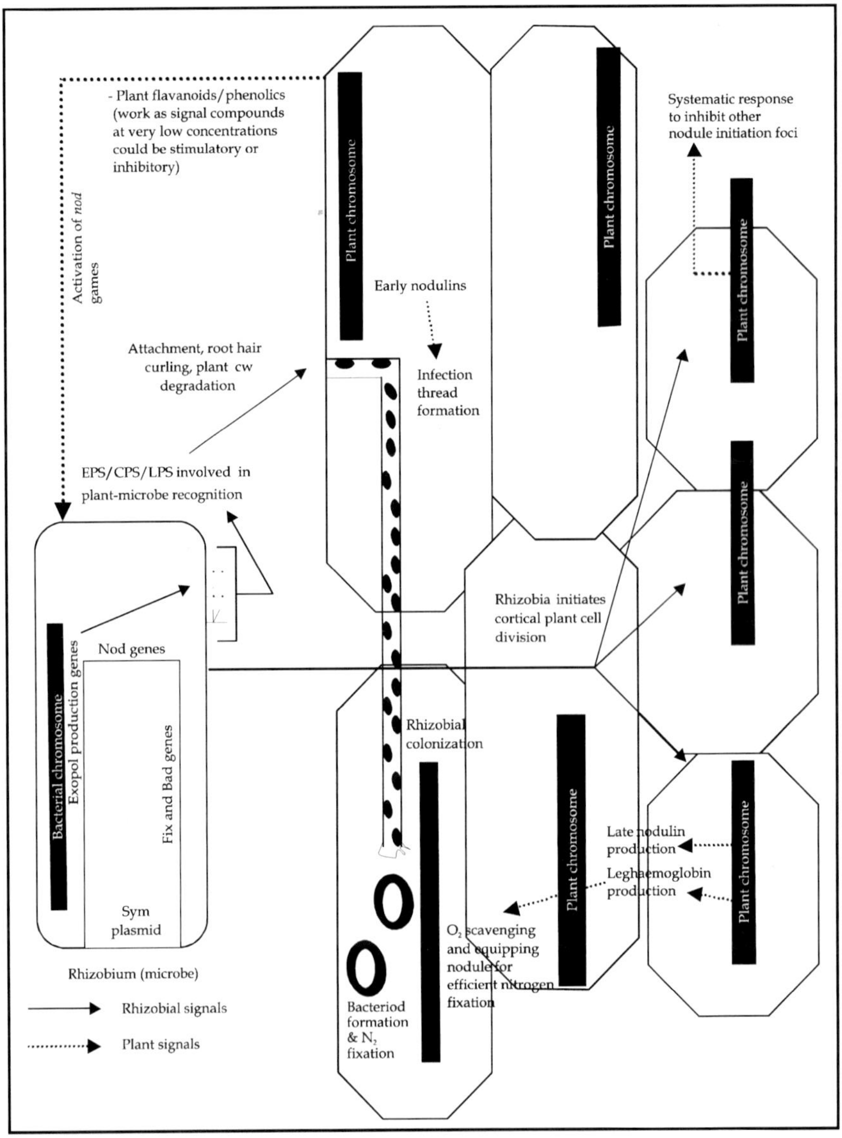

Host plant and rhizobial interactions for nodulation & nitrogen fixation

Extracellular proteins as microbial flagellin and plant differentiation factor (as CLAVATA 3) are the triggering factors for plant LRR receptor proteins (Spaink, 2002). Symbiosis receptor-like kinase (SYMRK) gene cloned from *Lotus* and pea are required in both arbuscular mycorrhizal and rhizobial recognition indicating convergent evolutionary lineage. It works in a symbiotic signal transduction pathway leading from the perception of microbial signal molecules to rapid symbiosis-related gene activation.

Genes regulating nodulation process are recessive and few in number as revealed by recent discovery of non-fixing mutants with altered patterns of nodulation (Tsyganov *et al*, 1998). Duc (1995) found that faba bean nodulation mutants including mutants with ineffective nodulation and supernodulation, had mutations of single recessive genes. There exists reports of presence of a common translocatable factor facilitating hypernodulation in various legume species in the shoots (e.g. in soybean, mung bean, *Vigna signali*, and hyacinth bean, *Lablab purpureus*, Harper *et al*, 1997). Godfroy *et al* (2006) have reported the presence of a calcium and calmodulin dependent protein kinase (CCaMK) coded by *dmi3* gene in *Medicago truncatula* involved in sensing/discriminating the nod factor regulated Ca^{2+} oscillations and further triggering of appropriate downstream signaling pathway.

b) Genetic manipulation of non-legumes

Recent developments in understanding the process of non-legume nodulation particularly rice has substantially led to conclusion that rice genome contains the early signal reception and nodule initiating proteins. Reddy *et al* (1999) reported presence of Enod gene homologues in *Oryza sp* followed by isolation of rice cDNA homologous to legume *enod40* and *enod93* genes (Kouchi *et al*, 1999). Moreover *Enod40* has been observed to play an important role in differentiation or functioning of vascular bundles (Kouchi *et al*, 2000). Incorporation of cDNA clone of *enod40* from *Medicago trunculata* in rice (Dey *et al*, 1990) caused activation of *Mtenod12* promoters in cortical cells, endodermis, pericycle and parenchymatous stele cells of rice roots upon treatment with NodNGR factor (10^{-6} to 10^{-9} M concentration) (Reddy *et al,* 1998).

c) Genetic manipulation of Rhizobium

The first *nif* genes identified by complementation were isolated from clone banks of the diazotroph *Klebsiella pneumoniae*. Studies with *K. pneumoniae* free living N-fixer have shown that the conversion of atmospheric nitrogen to ammonia requires at least 17 genes which are clustered on a single region of the chromosome.

The present scenario is flooded with reports of metabolic engineering of rhizobia for a variety of biochemical pathways (Bascon *et al,* 2000; Bhanu *et al,* 1994). Recently the *hup* genes that code for a hydrogenase system in *Rhizobium etli* cells were engineered to work under free-living conditions and the transformant cells showed a greater symbiotic efficiency and thus enhanced legume productivity. In another report a chimeric *nifHDK* operon regulated by the strong *nifHc* promoter sequence was constructed and expressed in symbiosis with the common bean plant (*Phaseolus vulgaris*) and comparison with the wild-type strain showed nitrogenase overexpression and increased nitrogenase activity, plant weight, nitrogen content in plants and higher seed yield, nitrogen content and nitrogen yield in seeds. Additionally, expression of the chimeric *nifHDK* operon in a poly-β-hydroxybutyrate-negative *R. etli* strain produced an additive effect in enhancing symbiosis (Peralta *et al,* 2004).

Brito *et al* (2002) replaced the native NifA-dependent *hupSL* promoter of *Rhizobium leguminosarum* bv. *viciae* UPM791 with the FnrN-dependent *fixN* promoter, generating strain SPF25, which expressed the *hup* system in microaerobic free-living cells. SPF25 reaches levels of hydrogenase activity in microaerobiosis similar to those induced in UPM791 bacteroids. Maximal hydrogen uptake activity was observed after 10 h of microaerobiosis. Immunoblot analysis of microaerobically induced SPF25 (pALPF1) cell fractions indicated that the HupL active form is located in the membrane, whereas the unprocessed protein remains in the soluble fraction. Symbiotic hydrogenase activity of strain SPF25 was not impaired by the promoter replacement. This offers a new method in increasing the hydrogenase activity under symbiotic conditions.

Rhizobial microsymbiont could also be made more competitive to get established in the rhizosphere (produce compounds inhibitory to *nod* gene expression in native rhizobia as gene for NolA protein - *nod* gene repressor) identified in USDA110 transferred to *B. japonicum* serocluster 123 caused depression of Nod factor production (Vlassak and Vanderleyden, 1997), out compete the native microflora (genes controlling catabolism of specific legume secondary metabolites like homoserine and trigonelline (Boivin *et al*, 1991), and exhibit greater nodulation (genes regulating nodule formation efficiency in *R. meliloti* (Sanjuan and Olivares, 1991) if they are genetically engineered.

Methods of estimation of nitrogenase activity in legume roots

Nitrogen contributed by BNF in plant body or soil is a question of considerable relevance and it demands quantification techniques that could assess the contribution of BNF. Several direct and indirect methods could be used to quantify or measure the nitrogen fixed by the activity of nitrogenase enzyme of the microsymbiont. Following two methods are most commonly used and provide approximate to advanced and accurate quantification of the N contributed by BNF are discussed below.

a) Acetylene reduction assay (ARA)

It is inexpensive, rapid, sensitive and apparently accurate method (Hardy *et al*, 1973) extensively used for field measurement of N_2 fixation. This assay gives an estimate of the activity of nitrogenase, an enzyme that reduces atmospheric nitrogen as well as several other compounds in nodules of host plant or under *in vitro* solution or semi solid medium conditions.

It is an indirect system of measuring nitrogen fixation in terms of reduction of acetylene (C_2H_2) to ethylene (C_2H_4). The basic assay involves incubation of the sampled detached nodules, nodulated root pieces, or detopped root system with 10% acetylene in a closed container of known volume for 45 minutes or one hour followed by analysis of the gas phase samples by gas chromatography to obtain the concentration of accumulated ethylene. It leads to measuring of electron flux through nitrogenase in the sample material under the conditions, which prevail during the assay period.

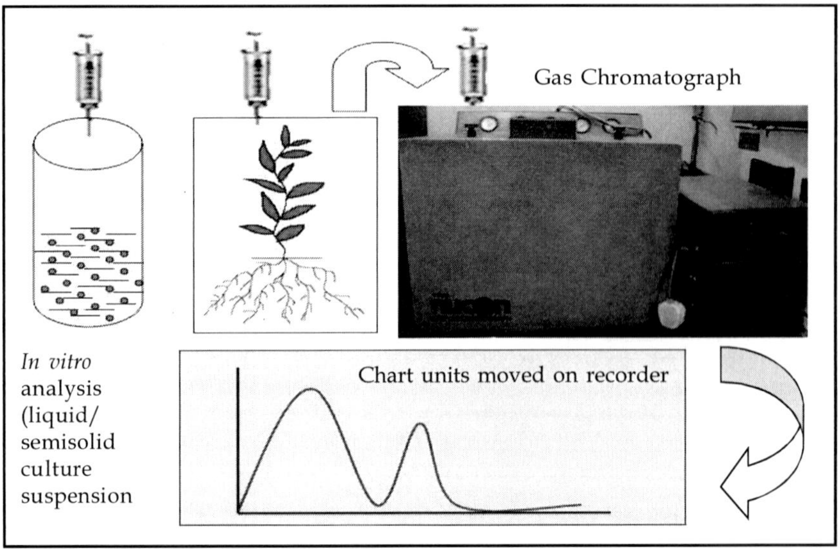

Acetylene reduction assay includes both *in vivo* (plant roots) and *in vitro* (liquid/semisolid bacterial culture) systems

b) ^{15}*N isotopic method*

This method is more accurate one as it helps in distinguishing between the plant N and the nitrogen fixed by the activity of nitrogenase enzyme system. Various methods are known for measuring stable isotope of nitrogen (^{15}N) with the most used one being isotopic dilution technique (McAulife *et al.*, 1958) that exploits differences in ^{15}N abundance of different N sources. The ^{15}N analysis is useful in obtaining

i) differences in ^{15}N of different N sources i.e. soil, atmosphere and fertilizer and thus could differentiate the plant N sources with more sensitivity and accuracy as well as provide integrated value for nitrogen fixation,
ii) enrichment of soil N due to direct use of ^{15}N method in the field situations
iii) exposure of N_2 fixing sites e.g. plant roots, to nitrogen and so could provide information simultaneously on the fertilizer use efficiency of both legume and non-legume reference crop
iv) could provide Ndfa proportion of even partially damaged plants
v) could help in ranking of genotypes for N fixing efficiency

Future Perspectives

Proteomic studies: Proteomics i.e. study of all the expressed proteins of a cell; is an ideal tool for the dissection of plant microbe interactions. It provides a broad overview of the proteins produced by both partners during their constant signal exchange and, in particular, it enables to ascertain the effect on gene product networks of gene knockouts, additions and specific growth states to be determined (Rolfe *et al,* 2003). It also allows the detection of signal transduction pathways by following phosphorylation changes of proteins that are important for protein function (Kalia and Gupta, 2005). The recent discovery of several plant receptor kinases responsible for the early detection and signal transduction of Nod factor perception (Endre *et al,* 2002) and autoregulation of nodule numbers (Nishimura *et al,* 2002), suggests that many early plant-microbe signaling events are regulated by phosphorylation events and key receptor kinases.

Non-legume nodulation: Now days the range of legume-rhizobia symbiosis has been extended to economically vital major food crops or cereals and certain rhizobia that are capable of crack entry from the corners of the ruptured epidermis during emergence of lateral rootlets in cereal crops (Kalia and Gupta, 2002) have been used for this purpose. However this process could be enhanced by use of adjunct procedures of prior hydrolytic enzyme treatment (Cocking *et al*, 1994), addition of phytohormones or use of signal chemicals (Kannaiyan *et al*, 2001).

References

Amutha K and Kannaiyan S (2000) Effect of kinetin and inoculation of *Azorhizobium caulinodans* on induction of lateral rootlets and paranodules in rice. *J Microb World*, 2:1-7.

Báscones E, Imperial J, Ruiz-Argüeso T and Palacios TM (2000) Generation of new hydrogen-recycling *Rhizobiaceae* strains by introduction of a novel *hup* minitransposon. *Appl Environ Microbiol*, 66:4292-4299.

Bejerinck MW (1888) Die Bacterien der Papilionaceen-Knöllchen. *Bot Zeitung*, 46:797-804.

Bejerinck MW (1890) Künstliche Infection von Vicia Faba mit *Bacillus radicicola*. Ernährungsbedingungen dieser Bacterie. *Bot Zeitung, 52:837-843.*

Bhanu N, Khanuja S and Lhoda M (1994) Integration of *hup* genes into the genome of chickpea-*Rhizobium* through site-specific recombination. *J Plant Biochem Biotechnol*, 3:19-24.

Boivin C, Barran L R, Malpica C A and Rosenberg C (1991) Genetic analysis of a region of the *Rhizobium meliloti* pSym plasmid specifying catabolism of trigonelline, a secondary metabolite present in legumes. *J Bacteriol*, 173:2809-2817.

Brito B, Palacios JM, Imperial J and Ruiz-Argüeso T (2002) Engineering the *Rhizobium leguminosarum* bv. Viciae Hydrogenase System for expression in free-living microaerobic cells and increased symbiotic hydrogenase activity. *Appl Environ Microbiol*, 68(5):2461-2467.

Burdon JJ, Gibson AH, Searle SD, Woods MJ and Brockwell J (1999) Variation in the effectiveness of symbiotic associations between native rhizobia and temperate Australian *Acacia*: Within-species interactions. *J Appl Ecol*, 36:398-408.

Cocking EC, Webster G, Batchelor CA and Davey MR (1994) Nodulation of non-legume crops. A new look. Agro Food Industry Hi-Tech, Jan/Feb, pp. 21-24.

De Lajudie P, Willems A, Pot B and 7 other authors (1994) Polyphasic taxonomy of rhizobia: emendation of the genus *Sinorhizobium* and description of *Sinorhizobium meliloti* comb. Nov., *Sinorhizobium saheli* sp. Nov., and *Sinorhizobium teranga* sp. Nov. *Int J Syst Bacteriol*, 44, 715–733.

Dey M, Torrizo LB, Chaudhari RK, Reddy PM, Ladha JK, Datta K and Datta SK (1999) Transgenic rice harbouring legume *ENOD40* gene. *Rice Genet Newsl*, 16:147-149.

Duc G (1995) Mutagenesis of faba bean (*Vicia faba* L.) and the identification of five different genes controlling no nodulation, ineffective nodulation or supernodulation. *Euphytica*, 83:147–152.

Duc G, Trouvelot A, Gianinazzi-Pearson V and Gianinazzi S (1989) First report of non-mycorrhizal plant mutants (Myc-) obtained in pea (*Pisum sativum* L.) and faba bean (*Vicia faba* L.). *Plant Sci.*, 60:215–222.

Endre G, Kereszt A, Kevei Z, Mihacea S, Kaló P and Kiss GB (2002) A receptor kinase gene regulating symbiotic nodule development. *Nature*, 417:962–966.

Fisher RF, Egelhoff TT, Mulligan JT and Long SR (1988) Specific binding of proteins from *Rhizobium meliloti* cell-free extracts containing NodD to DNA sequences upstream of inducible nodulation genes. *Genes Dev*,. 2:282-293

Frank B (1879) Über die Parasiten in den Wurzelanschwellungen der Papilionaceeen. *Bot Zeitung*, 24:377-388.

Fred EB, Baldwin IL and McCoy E (1932) Root nodule bacteria and leguminous plants. Univ. Wisconsin, Madison, WI, USA.

Fuchsius L (1542) De historia stirpium commentarii isignes. Michael Isingrin, Basel, Switzerland.

Godfroy O, Debelle F, Timmers T and Rosenberg C (2006) A rice Calcium- and Calmodulin-dependent protein kinase restores nodulation to a legume mutant. *Molecular Plant Microbial Interaction*, 19(5):495-501

Goodfellow M and O'Donnell AG (1993) The roots of bacterial systematics. In *Handbook of New Bacterial Systematics*, (eds) Goodfellow M and O'Donnell AG. London: Academic Press, pp.3-54.

Graham PH and Vance CP (2000) Nitrogen fixation in perspective: an overview of research and extension needs. *Field Crops Res.*, 65:93–106.

Hardy RWF, Burns RC and Holstein RD (1973) Applications of the acetylene ethylene assay for measurement of nitrogen fixation. *Soil Biol Biochem*, 5:47-81.

Harper JE, Corrigan KA, Barbera AC and Abd-Alla MH (1997) Hypernodulation of soybean, mung bean, and hyacinth bean is controlled by a common shoot signal. *Crop Sci.*, 37:1242–1246.

Hellriegel H (1886) Welche Stickstoffquellen stehen der Pflanze zu Gebote? *Landwirtsch Versuchstat Dresden*, 33:464-465.

Hellriegel H and Wilfarth H (1888) Untersuchungen über die Stickstoffnahrung der Gramineen und Leguminosen. Beilageheft zu der Zeitschrift des Vereins für die Rübenzucker-Industrie des Deutschen Reiches, Buchdruckerei der Post. Kayssler & Co., Berlin, Germany.

Hirsch AN, Lum MR and Downie JA (2001) What makes Rhizobia-legume symbiosis so special? *Plant Physiology*, 127:1481-1492.

Jarvis BDW, Downer HL and Young JPW (1992) Phylogeny of fast-growing soybean-nodulating rhizobia supports synonymy of *Sinorhizobium* and *Rhizobium* and assignment to *Rhizobium fredii*. *Int Syst Bacteriol*, 42: 93-96.

Kalia A and Gupta RP (2002) Nodule induction in non-legumes by rhizobia. *Ind J Microbiol*, 42:183-193.

Kalia Anu and Gupta RP (2005) Proteomics:A paradigm shift. *Critical Reviews Biotechnol*, 25:173-198.

Kannaiyan S, Kumar K and Gopalaswamy G (2001) Stem nodulating *Sesbania rostrata* as a bioresource for rice production, induction of nodulation and endophytic N_2 fixation in rice. *Madras Agric J*, 88(10-12):547-560.

Keefe RG and Maier RJ (1993) Purification and characterization of an O_2-utilizing cytochrome *c* oxidase complex from *Bradyrhizobium japonicum* bacteroid membranes. *Biochim Biophys Acta*, 1183:91-104.

Kouchi H, Takane K, So R, Ladha JK and Reddy PM (1999) Rice *ENOD40*:Isolation and expression analysis in rice and transgenic soybean root nodules. *Plant J*, 18:121-129.

Kouchi H, Takane K, So R, Reddy PM and Ladha JK (2000) Characterization of rice *ENOD40*:Do *ENOD40S* accomplish analogous functions in legumes and non-legumes? *In:The Quest for Nitrogen Fixation in Rice.* (eds) Ladha JK and Reddy PM, International Rice Research Institute, Philippines, pp. 263-272.

Krusell L, Madsen LH, Sato S, *et al*. 2002. Shoot control of root development and nodulation is mediated by a receptor kinase. *Nature*, 420: 422–426.

Malek W and Sajnaga E (1999) Current taxonomy of the rhizobia. *Acta Mirobiologica Polonica*, 48(2):109-122.

Malpighi M (1675) Anatome plantarum. Regiae Societati (J. Martyn), London, United Kingdom.

McAulife C, Chamblee DS, Uribe-Arango H and Woodhouse Jr. WW (1958) Influence of inorganic nitrogen on nitrogen fixation by legumes as revealed by ^{15}N. *Agron J*, 50: 334-337.

Nishimura R, Hayashi M, Wu GJ, *et al*. (2002) HAR1 mediates systemic regulation of symbiotic organ development. *Nature*, 420: 426–429.

Peralta H, Mora Y, Salazar E, Encarnación S, Palacios R and Mora J (2004) Engineering the *nifH* Promoter Region and Abolishing Poly-β-Hydroxybutyrate Accumulation in *Rhizobium etli* Enhance Nitrogen Fixation in Symbiosis with *Phaseolus vulgaris*. Appli. Environ. Microbiol., 70(6):3272-3281.

Perret X, Stahelin C and Broughton WJ (2000) Molecular basis of symbiotic promiscuity. *Microbiol Mol Biol Rev*, 64:180–201

Petersson B, Johansson KE and Uhlen M (1994) Sequence analysis of 16s rRNA from mycoplasmas by direct soild-phase DNA sequencing. *Appl Environ Microbiol*, 44:374.

Postgate J (1998) Nitrogen Fixation, Cambridge University Press, Cambridge, United Kingdom.

Prazmowski A (1889) Das Wesen und die biologische Bedeutung der Wurzelknöllchen der Erbse. *Bot Centbl*, 39:356-362.

Prazmowski A (1890) Die Wurzelknöllchen der Erbse. *Landwirtsch Versuchsstat Berlin*, 37:161-238.

Preisig O, Anthamatten D and Hennecke H (1993) Genes for a microaerobically induced oxidase complex in *Bradyrhizobium japonicum* are essential for a nitrogen-fixing endosymbiosis. *Proc Natl Acad Sci USA*, 90:3309–3313.

Reddy PM, Ladha JK, Ramos MC, Maillet F, Hernandez RJ, Torrizo LB, Oliva NP, Datta SK and Datta K (1998) Rhizobial lipochitooligosaccharide nodulation factors activate expression of the legume early nodulin gene *enod12* in rice. *Plant J*, 14:693-702.

Reddy PM, Ladha JK, So R, Hernandez RJ, Ramos MC, Angeles OR, Dazzo FB and deBruijn FJ (1997) Rhizobial communication with rice roots:induction of phenotypic changes, mode of invasion and extenet of colonization. *Plant Soil*, 194:81-98.

Rolfe BG, Mathesius U, Djordjevic M, Weinman J, Hocart C, Weiller G and Bauer WD (2003) Proteomic analysis of legume–microbe interactions. *Comp Funct Genom*, 4:225-228.

Ruiz-Argüeso T, Palacios JM and Imperial J (2000) Uptake hydrogenases in root nodule bacteria, p. 489-507. *In* E. W. Triplett (ed.), Prokaryotic nitrogen fixation. A model system for the analysis of a biological process. Horizon Scientific Press, Norfolk, England.

Sahgal M and Johri BN (2003) The changing face of rhizobial systematics. *Curr Sci* 84(1):43-48.

Sanjuan J and Olivares J (1991) *Nifa-NtrA* regulatory system activates transcription of *nfe*, a gene locus involved in nodulation competitiveness of *Rhizobium meliloti*. *Arch Microbiol*, 155:543-548.

Sawada H, Kuykendall LD and Young JM (2003) Changing concepts in the systematics of bacterial nitrogen-fixing legume symbionts. *J Gen Appl Microbiol*, 49, 155-179.

Spaink HP (2002) A receptor in symbiotic dialogue. *Nature*, 417, 910–911

Stracke S, Kistner C, Yoshida S, Mulder L, Sato S, Kaneko T, Tabata S, Sandal N, Stougaard J, Szczyglowski K and Parniske M (2002) A plant receptor-like kinase required for both bacterial and fungal symbiosis. *Nature*, 417:959–962.

Sy AE, Giraud P, Jourand N, Garcia A, Willems P, de Lajudie Y, Prin M, Neyra M, Gillis C, Boivin-Masson and Dreyfus B (2001) Methylotrophic *Methylobacterium* bacteria nodulate and fix nitrogen in symbiosis with legumes. *J Bacteriol*, 183:214–220.

Terefework ZG, Lortet L, Suominen and Lindstrom K (2000) Molecular evolution of interactions between rhizobia and their legume hosts. *In* E. W. Triplett (ed.), *Prokaryotic nitrogen fixation: A model system for the analysis of a biological process*, pp. 187–206. Horizon Scientific Press, Wymondham, UK.

Thies JE, Holmes EM and Vachot A (2001) Application of molecular techniques to studies in *Rhizobium* ecology: a review. Australian *J Expt Agricul*, 41(3):299–319. Full text doi:10.1071/EA99171

Thomas RJ, Askawa NM, Rondon MA and Alarcon HF (1997) Nitrogen fixation by three tropical forage legumes in an acid soil savanna of Colombia. *Soil Biol Biochem*, 29:801-808.

Tsyganov VE, Morzhina EV, Stefanov SY, Borisov AY, Lebsky VK and Tikhonovich IA (1998) The pea (*Pisum sativum* L.) genes *sym33* and *sym40* control infection thread formation and root nodule function. *Mol Gen Genet*, 259:491–503.

Vij N (2003) Proteomics: a novel approach to explore signal exchanges in *Rhizobium*-legume symbiosis. Indian *J Exp Biol*, 41(10):1133-5.

Viprey V, Rosenthal A, Broughton WJ and Perret X (2000) Genetic snapshots of the *Rhizobium* species NGR234 genome. *Genome Biol*, 1(6) research:0014.1-0014.17.

Vlassak KM and Vanderleyden J (1997) Factors influencing nodule occupancy by inoculant rhizobia. *Crit Rev Plant Sci*, 16:163-229.

Wang ET and Martinez-Romero E (2000) Phylogeny of root- and stem-nodule bacteria associated with legumes. *In* E. W. Triplett (ed.), *Prokaryotic nitrogen fixation: A model system for the analysis of a biological process*, pp. 177–186. Horizon Scientific Press, Wymondham, UK.

Waters JK and Emerich DW (2000) Transport of metabolites to and from symbiosomes and bacteroids. *In* E. W. Triplett (ed.), *Prokaryotic nitrogen fixation: A model system for the analysis of a biological process*, pp. 549–558. Horizon Scientific Press, Wymondham, UK.

Wayne LG, Brenner DJ, Colwell RR, Grimont PAD, Kandler O, Krichevsky L, Moore LH, Moore W, Kandler O, Krichevsky L, Moore LH, Moore W and Trüper HG (1987) Report of the ad hoc committee on reconciliation of approaches to bacterial systematics. *Int J Syst Bacteriol*, 37:463-464.

Willems A and Collins MD (1993) Phylogenetic analysis of rhizobia and agrobacteria based on 16S rRNA gene sequences. *Int J Syst Bacteriol*, 43 (2):305–313

Woronin MS (1866) Über die bei der Schwarzerle (*Alnus glutinosa*) und bei der gewöhnlichen Gartenlupine (*Lupinus mutabilis*) auftretenden Wurzelanschwellungen. Mémoires de l'Academie Impériale des Sciences de St. Pétersbourg, VII Series, vol. X.

Yang D, Oyaizu Y, Oyaizu H, Olsen GJ and Woese CR (1985) Mitochondrial origins. *Proc Natl Acad Sci USA*, 82:4443–4447.

Young JPW (1996) Phylogeny and taxonomy of Rhizobia. *Plant Soil*, 186: 45-52.

Mind bogglers

1. Why rhizobial taxonomy is constantly being revised and isolates are redistributed to form new genera/species/subspecies/varieties?
2. Why rhizobia exhibit symbiosis in legumes and under what conditions?
3. What are the functions of the *fix* genes in imparting and continuing symbiosis? Do free living nitrogen fixers have *fix* genes and why?
4. What are the physiological and metabolic changes in rhizobia and host cells that lead to formation of a conducive environment for N_2 fixation?
5. Enlist and explain various methods for the estimation of nitrogen incorporated in plant body by biological nitrogen fixation by rhizobia. Which method do you suggest to be most appropriate for routine research work?
6. Justify the statement "*nod* genes code for proteins (invariably enzymes)". What is type of compound

synthesized by the activity of nod gene products? Also mention the role of the compound thus synthesized.

7. The future of rhizobial systematics largely relies on the application of novel molecular biological tools and techniques.
8. Why researchers are now more inclined in deciphering the proteomic changes in both macro (plant host) and microsymbiont (rhizobia) during initiation and creation of symbiosis? How it is helpful in monitoring plant-microbial endosymbiosis?
9. What is the feasibility, benefit and applicability of use of genetically modified host plants or the rhizobia?
10. Comment on the statement 'Nitrogen fixation is a high energy demanding process'.
11. Legume-rhizobial symbiosis involves very intricate and finely tuned choreography of an array of signals. What are the signals and illustrate their types? How these are helpful in determining host specificity for rhizobia?
12. Would you agree with the statement that 'The answers to the queries of extension of symbiosis to non legumes could be obtained more easily by molecular techniques as genomics and proteomics" and why?

2

Actinomycetes Symbiosis with Higher Woody Plants

- Introduction 38
- Classification of *Frankia* *39*
- Ecology of *Frankia* 42
- Symbiosis with Higher Woody Plants 47
- Future Perspectives 54
- References 55
- Mind Bogglers 60

INTRODUCTION

Rhizobia and *Frankia* are soil bacteria that exhibit symbiosis with their particular hosts owing to their ability to understand the intricate signals excreted by the host roots. Rhizobia exhibit symbiosis with legumes and one species member of Ulmaceae family *Parasponia andersonii* while actinomycetes *Frankia* forms the group of nitrogen-fixing bacteria that live in symbiosis with a large spectrum of dicot plants distributed among eight families consisting of over 24 genera and around 200 species of angiosperms (Benson and Silvester, 1993) and are collectively responsible for 15% of the biologically fixed nitrogen in the world. This *Frankia*-woody tree or dicotyledonous plant interaction results in evolution of the actinorhizal plants and the symbiotic interaction is described between *Frankia* (actino) and plant roots (rhiza).

Actinorhizal plants have a worldwide distribution and represent mostly perennial, dicotyledonous woody shrubs and trees. They are commonly used in forestry as fuelwood, pulpwood or saw timber. Due to their ability to enter into symbiosis with N_2-fixing *Frankia*, actinorhizal plants are successful pioneer species frequently found in low-nitrogen sites and sites/area after disturbances such as fires, volcanic eruptions and floodings that are unfavourable for other N_2-dependent plant species. Furthermore, a well-developed actinorhizal plant root system favors soil-binding, which improves the quality of impoverished soils and strongly supports the use of these plants in land reclamation. Actinorhizal plants are also commonly used as intercrops for other tree species. The contribution to nitrogen fixation in the *Frankia*-alder symbiosis can be up to 300 kg/ha/yr (Tarrant and Trappe, 1971). Although the nitrogen-fixing potential of alder-*Frankia* symbiosis may be high, the amount of nitrogen actually fixed in the field is often low because of unfavorable environmental conditions. Therefore, proper management practices that optimize efficiency of the nitrogen-fixing system are required (Dommergues, 1997). Certain actinorhizal plant species include *Alnus, Hippophae, Myrica, Shepherdia, Elaeagnus, Ceanothus, Purshia* and *Dryas*.

CLASSIFICATION OF FRANKIA

The Actinomycete genus *Frankia* belongs to the recently emended family, Frankiaceae (Normand, 1996) including Gram positive bacteria, able to establish a nitrogen-fixing symbiosis with roots of angiospermous woody plants. This association is most vocal and prominent in alder (*Alnus spp.*) and myrtle (*Myrica spp.*) that grow in nitrogen limiting temperate regions as forest clearings, mine wastes, sand dunes and glacial moraines.

Typical *Frankial* strains could be classified and identified as *Frankia* using morphological, cytochemical and physiological criteria. Morphological criteria include presence of Gram-positive filamentous bacteria that exhibit formation of septate, branching hyphae, production of multilocular sporangia, presence of non-motile spores in multilocular sporangia and production of thick-walled, lipid-encapsulated structures called vesicles, which are the site for nitrogen fixation (Meesters *et al*, 1987). Morphologically *Frankia alni* causes root hair deformation, it penetrates the cortical cells and induces the formation of nodules, which resemble those induced by *Rhizobium* in legumes. These nodules are then colonized by vegetative hyphae (mycelium filaments), which differentiate into diazo-vesicles. It is in these specialized, spherical, thick-walled cells that reductive nitrogen fixation takes place, protected from molecular oxygen by numerous layers of tightly stacked hopanoid lipids. Most *Frankia* isolates can be grouped into four major host specificity groups though this classification is incomplete due to difficulty in isolation of *Frankia* isolates from all actinorrhizal plants (Table 1).

Host specificity has been the criteria of choice for classifying these actinomycetale members but other methods *viz*

Table 1. Groups of *Frankia* on basis of host specificity.

Groups	Major hosts nodulated	Signature feature	Reference
I	*Alnus, Comptonia* and *Myrica*	Highly promiscuous, do not nodulate members of Coriaceae, Datiscaceae, Rhamnaceae and Rosaceae	(Baker, 1987)
II	*Casuarina* and *Myrica*	Highly host specific	-do-
III	*Myrica* and *Eleagnus*	Lack N-acetyl glucosamine and sialic acid binding lectins on cell surface	-do-
IV	Members of the Elaeagnaceae (*Elaeagnus, Hippophaë, Shepherdia*)	Nodulates homologous host species	-do-
Atypical *Frankia*	Non-infective strains, Which are unable to form root nodules on host plants	Differ in morphological, cytochemical and physiological characters than normal frankiae isolates	(Bosco *et al*, 1992).
Inefficient isolates	Nodulate variety of hosts form inefficient nodules	Unable to fix nitrogen under pure culture or symbiosis	-do-
Broad host range isolates	Infect variety of host plants	Fix nitrogen under pure culture or symbiosis	-do-

morphological, physiological, and nucleic acid based criteria have been proposed for the classification of members within the genus *Frankia*. The various methods used for obtaining a coherent classification at the species level included soluble protein pattern analysis, isoenzyme patterns, fatty acids analysis, serological studies, DNA-DNA relatedness and genome and plasmid restriction analysis (Huguet *et al,* 2001). However no standardized criteria for *Frankia* species definition have been established (Lechevalier, 1994).

The %G+C studies revealed that this group of bacteria have higher %G+C amounts than the one predicted for the nitrogen fixing proteobacteria including rhizobia and free living azospirilla. The DNA sequence based protocols using various DNA templates/sequences could help clarify the provision of subgroups among various host specificity groups. Based on comparative sequence analysis of 16S rDNA sequences of *Frankia* pure cultures and uncultured endophytes in nodules, *Frankia* was considered to be the only genus within the family *Frankiaceae* (Normand *et al,* 1996). Following four main subdivisions have been proposed within the genus *Frankia* by Normand *et al* (1996):

- *Frankia alni* and other typical nitrogen fixing strains belonging to the *Alnus* and *Casuarina* host infection groups,
- uncultured symbionts of *Dryas, Coriaria* and *Datisca* species,
- *Elaeagnus* infective strains and
- atypical non-nitrogen-fixing strains

The 16S rDNA variability among *Frankia* is low and phylogenetic relationships are sometimes ambiguous with partial sequences or with closely related strains (Benson and Clawson, 2000). Thus using the more variable *nifH* gene allows more detailed analysis in only nitrogen fixing strains. Now days the variability in conserved sequences of glutamine synthetase

I (*glnA*) gene has been harnessed due to higher variability than 16S rRNA gene & almost universal presence of the former gene in prokaryotes. Above all *glnA* gene could be used for both the nitrogen fixer and also non nitrogen fixing strains and it experiences different selective pressure than either *nifH* or 16S rDNA (Clawson *et al*, 2004). Benson and Clawson (2000) have reported the grouping of effective *Frankia* strains on basis of 16S rDNA or *nifH* sequence analysis into three major clades and further their subgrouping and identification on basis of behavior in culture, morphology & mode of infection (Table 2).

Another comparative sequence analysis based on PCR amplification and cloning of an actinomycetes-specific 23S rRNA insertion mostly agreed with the former differentiation and grouped 35 *Frankia* strains into roughly seven different host infection groups. This comparative analysis demonstrated that strains belonging to *Casuarina* (HIG II) and *Elaeagnus* host infection groups (HIG VI) showed limited sequence variation, whereas *Alnus* group could be separated into four different subgroups (HIG I, IIIa, IIIb, and IV), three (IIIa, IIIb and IV) containing typical nitrogen-fixing strains and a fourth (I) containing only non nitrogen-fixing strains (Honerlage *et al*, 1994).

Ecology of Frankia

Soil *Frankia* population is very less as these bacteria are nearly impossible to be isolated in pure culture from soil in presence of other native and common soil bacterial populations. Very few have reported isolation of *Frankia* from soil. This problem is inherent due to long generation times, the requirement for special isolation factors (Quispel *et al*, 1989) and the lack of specific nutritional requirements of pure cultures (Akkermans *et al*, 1992). There has been no universal, standardized isolation protocol for isolation of *Frankia* from soil or root nodules of actinorrhizal plants.

- first *Frankia* isolated (Berry and Torrey, 1979)
- several hundred *Frankia* strains now been identified and isolated

- all three *Frankia* clades emerged around 95 Myr bp well before appearance of angiosperms (Cretaceous period 125 Myr bp) (Crane *et al*, 1995) and before members of the higher hamamelids (Sims *et al*, 1999)

Soil *Frankia* populations could be identified using plant bioassays that quantify the nodulation capacity on a specific host plant providing the infective *Frankia* population. The nodulation capacity (expressed as nodulation units g^{-1} soil) of a *Frankia* source is defined as the number of *Frankia* particles per unit of *Frankia* source, each of which induces one nodule in a nodulation test. The quantitative methods include capture plants generally *Alder* species that get nodulated by soil *Frankia* isolates and the nodule so formed could provide the infective *Frankia* isolates. The basic protocol followed includes serial dilution of soil samples and inoculation of capture plant by serially diluted soil sample. The MPN method is used to score nodulation as +/- and the number of nodulation units estimated based on MPN statistics (Huss-Danell and Myrold, 1994).

The *Frankia-Alnus* association is a unique microbial interaction with higher plants that include a variety of physiological capabilities of both plant species and the microsymbiont. The age of plant (indicative of its active physiological status) determine the infection by the probable infective actinomycetes particles in rhizosphere soil while the types or variety of native rhizosphere microbial population as well as the number of frankiae and age of nodule also determines the extent of symbiotic benefit (Verghese and Misra, 2002). Thus the physiological activity of the two symbiotic partners has a role in determining the functional compatibility, which is prerequisite for phenotypic expression of the frankial association (frankial effectiveness). The compatibility depends on both the genetics of *Frankia* (lower *in situ* than *in vitro* nitrogen fixing rates of pure culture) and host plant (regulates nodulation by secretion of at least two specific signals) as well as the abiotic factors that govern the extent of effective

Table 2. *Frankia* isolates grouped on basis of 16s rRNA sequence analysis (Phylogenetic analysis)

Groups (Strains)	**Behaviour in culture**	**Infection of host**	**Major host**	**Special characters**	**Special structures for symbiosis**
Clade I	Not isolated in culture	infect hosts by intercellular penetration through the root surface	Members of Rosaceae genera *Cercocarpus, Chamaebatia* and Purshia and unisolated strains of *Datisca, Coriaria* and *Ceanothus*	-Not deciphered	form primitive vesicles in symbiosis
Clade II	Culture obtained	• Infects host (hamamelids) by root hair infection • Loss of ability to infect host by intercellular penetration	Higher hamamelid members	grow on organic acids	Form multicellular lipid enveloped vesicles in both culture and symbiosis
Clade III	Culture obtained	Infects host by root hair penetration	Members of Utricales, family Elaeagnaceae, *Ceanothus americanus, Purshia tridentata*	grow on both organic acids and simple sugars	Form multicellular lipid enveloped vesicles in both culture and symbiosis

symbiosis. These abiotic factors include pH, soil, soluble phosphorus, calcium levels, available nitrogen or nitrate concentration of soil, moisture, conductivity of water, climate, light availability, canopy cover, etc that bring about imbalance in the symbiotic relationship observed as significant variations in the nitrogen fixation capacities of the isolates (Han and New, 1998).

Bulloch (1994) reported artificial inoculation with *Frankia* culture *HFPCcli* to cause nodulation in *Casuarina cunninghamiana* and *C. glauca* plants in New Zealand and also reported in their long term study (after 6 to 8 years of inoculation) that this artificial inoculation resulted in significant increase in stem diameter and height (equivalent to 16-61% additional volume) and inoculated plants exhibited higher vigour over trees not nodulated at planting. The microbial ecological studies exhibited strain specific interaction responses of *Frankia* with other microbes (bacteria and fungi particularly VAM fungi) present in the rhizospheric niche regions of woody plants. Yamanaka *et al* (2002) have reported that the tripartite associations among *Frankia, Alpova diplophloeus* (an ectomycorrhizal fungus) and *Alnus tenuifoliain* improves the growth, nitrogen fixation, ectomycorrhizal formation, and mineral acquisition of *A. tenuifolia* observed as increased mineral content (P, K, Ca, Fe, Mg, Mn, Na, Si and Al) in the seedlings after 6 months of growth. Though positive response is observed in terms of increased biomass and yield of trees undergone dual inoculations (*Frankia* and AMF), the extent of benefit depends upon specific strains (Wheeler *et al,* 2000). In case of positive responsive strains there occurs increase in nodule biomass as well as number owing to enhanced P uptake and availability of essential micronutrients for smooth working of nitrogenase system in nodules.

- *Gymnostoma nodiflorum* and *G. deplancheanum* form small (<500mm long) and numerous spherical or pyriform mycorrhizal nodules
- mycorrhizal nodules arranged in two or three close rows along the side of the roots

- their number of increases with age.
- 300-mm wide root of *G. deplancheanum* can bear up to 120 mycorrhizal nodules cm^{-1}.
- white coloured during young stage become brown and surrounded with tannins on maturity
- epidermis of mycorrhizal nodules never form root hairs.
- nodules die within 3–4 months after their full development and get detached from root (Douhoux *et al,* 2001)

Molecular ecological techniques for detection of Frankia

With the advent of new molecular ecological techniques for isolation, detection and enumeration: novel bacteria, bacteria requiring fastidious/uncharacterized nutrients/components and certain bacteria occurring in very low numbers as *Frankia* could be detected easily thus helping in enriching our information regarding their unique roles in specialized and native ecological niches. These methods are based on detecting the presence of particular molecules particularly macromolecules as certain DNA stretches, specific protein markers, carbohydrate or lipid molecules from a given soil sample. The popularity of a molecular marker resides in its efficiency with which it is able to predict presence or advocates the absence of a particular microbe(s) in a given niche. The most popular molecular markers, such as the rRNA-encoding genes, contain conserved and highly variable areas that provide good targets for the detection, identification and phylogeny of cultured and uncultured microorganisms at different levels of resolution.

Hahn *et al* (1999) have summarized the role of molecular biological techniques in direct detection of *Frankia* populations from soil and nodules. Techniques used for analysis of *Frankia* populations are based on nucleic acid hybridization using 16S rRNA- and 23S rRNA sequences as markers (Clawson *et al,* 1999). Usually PCR-amplified partial 16S rRNA sequences from isolates or uncultured *Frankia* populations in root nodules are analyzed. Furthermore, specific *Frankia* populations can be analyzed by probing PCR products with specific oligonucleotide probes or primers directed against the 16S rRNA gene in dot blots or in PCR assays. The *Frankia* isolates could also be detected *in situ* using various *in situ* hybridization techniques that includes use of a range of probes as digoxigenin-, fluorochrome-labelled or enzyme labeled oligonucleotide probes (Zepp *et al*, 1997a; Zepp *et al*, 1997b). Oakely *et al* (2004) have sampled eight *Caenothus* species from six major biogeographic regions in and around California for developing a quantitative model for testing relative importance of geographic separation, host specificity, and environment in influencing the identity of *Ceanothus Frankia* symbionts as determined by rDNA sequence data. They reported that *Frankia* strains symbiotic with *Ceanothus* exhibited a high degree of genetic similarity.

The *in situ*, or whole cell hybridization, technique is based on the microscopical detection of naturally amplified target molecules like ribosomal RNAs. The targets are detected after the formation of stable hybrids with respective types of probes in fixed cells. These molecular markers could detect presence of *Frankia* strains from alder root nodules, aquatic systems, sediments and soils.

Symbiosis with Higher Woody Plants

Nodule formation (Frankia as an endophyte): *Frankia* populations inhabit three distinct ecological niches, the root nodules, the rhizosphere, and the soil, where *Frankia* can be found as a symbiont of actinorhizal plants, an associate of non-host plants or as a saprophyte, respectively. Although the biochemical and molecular basis of *Frankia*-actinorhizal plant symbiosis are not as well understood as in the case of counterpart *Rhizobium*-legume symbiosis, there occurs regulated series of events leading to this close association between infective *Frankia* and compatible host plant and the subsequent formation of root nodules on host plant roots.

Events of nodule formation:

i) *Infection process:* The initial events of actinorhizal symbiosis have remained obscure and still our knowledge of the nature and amount of the early signaling chemicals in *Frankia*-actinorhizal plants is marginal. The microscopic studies (Wall, 2000) and molecular biological studies have cleared the steps of infection process, nodule formation and colonization by *Frankia* (Guan 1997). Tunlid *et al* (1989) have reported difference in the fatty acid composition of the frankial wall during free living and in symbiosis with the host plant. Reports regarding induction of root hair deformation in host plant roots due to presence of few factors (speculated to be nod factor-like counterparts in *Frankia* showing positive inducible response towards flavanoid extracts) in culture supernatant of *Frankia* have provided a firm ground for intricate signaling events during this symbiosis. However the activity of the frankial factors is exhibited at high concentration (10^{-5}M) unlike the nod factor in rhizobia. *Frankia* in actinorhizal plants infect following root hair infection process (via deformed root hair as in *Alnus,* members of *Casuarinaceae* and *Myricaceae* family where *Frankia* induces cell wall hydrolysis or exhibits direct penetration of the root through intercellular spaces of epidermis and cortex as in Elaeagnaceae and *Ceanothus*) to form nodules having lateral root like internal anatomy (i.e. presence of cortical vascular bundle cylinder and periderm) (Huss-Danell, 1997). The cortical tissue of the nodule is inhabited by vegetative mycelial state of bacteria.

Structural zones of actinorhizal nodules

- Zone I: formed by the meristem,
- Zone II (prefixation zone): contains cells that become infected
- Zone III (fixation zone): Cels involved in Nitrogen Fixation
- Zone IV (senescence zone): degradation of plant cytoplasm and bacteria
- Interzone II-III: absent in actinorhizal nodules (Yang *et al,* 1991)

The frankial penetration initiates a series of cell division reactions in the hypodermis and cortex of the root to form a prenodule, which is unique as it consists of both infected and uninfected cells. The uninfected cells are the site of starch storage however the infected zone of legume and non-legume nodules by rhizobia contains only one type of cell population. Being a symbiotic association the host cell encapsulates the symbiont (*Frankia* hyphae) by production of pectin, hemicellulose and cellulose (wall-like material) which not only avoids defense response against the invading frankial hyphae (Berg, 1990) but also help in exchange of metabolites between the symbiotic partners (Smith and Smith, 1990). The actinomycetal nodules are multiple lobed having determinate meristem, and are perennial having morphologically differentiated clusters of lateral roots. Being a nitrogen fixer *Frankia* produces nitrogenase enzyme in vesicles at the end of its hyphae under free-living state as the enzyme complex needs oxygen protection by multilayered lipid envelope (Berry, 1994).

Altered host cell cytology on frankial infection

- ❒ cytoplasm undergoes important modifications
- ❒ increased metabolic and golgi activities
- ❒ highly polyploid nucleus
- ❒ greatly enlarged nucleolus
- ❒ increased endoplasmic reticulum

Molecular biology of nodule formation: In actinorhizal plants, nodules are derivatives of lateral root primordia and proteins analogous to legume nodulins may not exist. Goetting-Minesky and Mullin (1994) have shown that a nodule-specific transcript of host plant origin is present in total RNA of *Alnus glutinosa*. The amino acid sequence analysis exhibits similarity with cystein proteinase speculated to have role in defense response to *Frankia* invasion. Differential screening of *A. glutinosa* nodule cDNA library revealed that a subtilin-like protease gene expression is enhanced during early stages of nodule development in the infected cells before the onset of nitrogen fixation (Laplaze *et al*, 2000). Nodule-specific clones, *agNt84* and *ag164* which encodes glycine and histidine-rich proteins have also been isolated from *A. glutinosa* cDNA libraries constructed from root nodule mRNA. *In situ* hybridization shows that this gene is expressed only in infected cells of the prefixing zone of the young nodules (Pawlowski *et al*, 1997). The encoded proteins share sequence similarity with nodulin-24, a protein localized in peribacteroid membrane. In *Casuarina glauca-Frankia* symbiosis, chalcone synthase transcripts were isolated from nodule cDNA library. Chalcone synthases are present in nodules, roots and aerial organs. This enzyme could contribute to the establishment of the symbiosis and nodule development (Gherbi, 1996).

- novel actinomycetal primary (housekeeping) σ^{70} factor identified by screening of a *Frankia* strain EaI-12 gene library
- phylogenetic analysis classified σ^{70} in RpoD cluster of proteobacteria and low-G+C content gram-positive factors

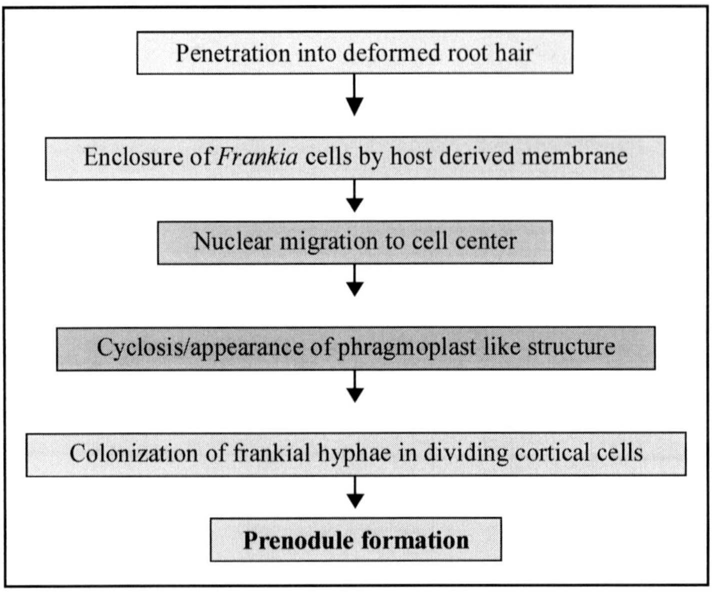

Formation of prenodule in actinorhizal plants by the intracellular invasion process followed by *Frankia*

- total DNA blots and PCR screening detected one or two *rpoD*-like DNA regions per species.
- *rpoD* matched the conserved region in all of the species tested
- *sigA* appeared to be strictly distributed among *Frankia* species infecting plants by the root hair infection process. (Lavier *et al,* 2004)

Nitrogen fixation in nodules by Frankia: Being a heterotroph the frankial symbiosis in the host plant roots depend on host photosynthate or carbon compounds for carrying out of nitrogen fixation which is an energy intensive process. The exact nature of compounds furnished by the host is not clear, as few *Frankia* strains are not able to use simple sugars in pure culture owing to absence of glycolytic enzymes. Hence it has been proposed that these microbes may be utilizing host lipids as carbon source as it has been reported by Selim *et al* (1996) that the growth of bacteria increased by addition of long-chain fatty acids to the growth medium. Nitrogen reduction during nitrogenase action results in production and release of hydrogen due to proton reduction, which causes a decline in the yield. Recycling of the hydrogen formed by nitrogenase catalyzed reduction of atmospheric nitrogen could be carried out by uptake hydrogenase and also this may help in oxidizing atmospheric hydrogen in the soil, thus increasing the energy resources of the bacterium. Gentili *et al* (2006) have reported the role of externally applied P and N fertilizers on the frankial-actinorhizal association and recorded that P at a medium concentration (0·1 mM) stimulated cell divisions in the cortex and number of prenodules, nodule primordia and emerging nodules as compared to higher or lower P levels while a high N level inhibited early cell divisions in the cortex, and this was particularly evident when the length of cell division areas and presence of the nodulation stages were related to root length.

Genetics of actinorhizal plants: The actinorhizal plants have evolved several novel genes like legumes to harbour the frankial association. Using molecular techniques, actinorhizal nodulin genes were found to be expressed in particular cell types. Majority of actinorhizal plants exhibit expression of various genes and accumulation of protein or other products for ease and finesse in working of nitrogenase system of the microsymbiont *Frankia*. A gene *cg12* from *Casuarina glauca* has been cloned and characterized recently that was observed to share strong homology to subtilisin-like protease gene families of several plants, including the actinorhizal nodulin gene *ag12* of *Alnus glutinosa*. This gene expresses early during nodule development just before onset of nitrogen fixation representing an early actinorhizal nodulin gene and may have a role in processing of extracellular or cell wall proteins during early stages of *Frankia* infection. Other early actinorhizal genes such as *ag164* and *agNt84* have been characterized in *Alnus glutinosa* that code for Ag164 and AgNt84 proteins, which are speculated to be involved in transport/storage of metal during the infection process (Laplozae *et al.*, 2000).

Actinorhizal root nodules like legume nodules contain haemoglobin (high amounts in *Casuarina* and *Myrica*, lower amount in *Alnus* and *Elaeagnus* and absent in *Datisca* spp) having sequence similarity with few *Rhizobium* nodulated plants. However in *Datica* subernization of cell walls surrounding the microbe is carried by the plant, which restricts the access of oxygen to *Frankia* cells. Apart from production of oxygen scavenging haemoglobin protein actinorhizal plants also follow few other techniques as production of nodule-specific glutamine synthase and transportation of asparagine or glutamine or citrulline (*Alnus*) in the nodules.

Genetics of Frankia: The genetic studies of *Frankia* have been lacking behind due to several hampering factors as slow growth rates, multicellular nature of the filament/hyphae, poor germination and lack of genetic markers. Now that the molecular/genetic markers have been isolated and identified, such as the antibiotic resistance patterns of the frankial isolates (Tisa *et al*, 1999), heavy metal resistance markers (Richards *et al*, 2002), these could be used for screening their presence in soil, rhizospheric as well as nodule niches and would prove beneficial in establishment studies.

The *Frankia* shares little homology with legume symbiont rhizobia regarding nodulation and nitrogen fixation genes. Unlike rhizobia frankial nod genes are present on the chromosome and the difference exists in the invasion process too. The frankial nod genes include nod ABC genes responsible for production of factor Rm1 (derivative of N-acyl-β-1,4-D-glucosamine tetrasaccharide), nodD (nod gene elicitor), nodSUIJ, nodY, nodK (involved in production of enzymes involved in substitutions on basic nod factor strucure) etc. In *Frankia* invasion process may involve degradation of pectin due to production of pectate lyase enzyme coded by *pel* genes along with the cellulolytic activity (Verghese and Misra, 2002). Presence of *hup* genes in *Frankia* has also been reported, but sequence homology with other nitrogen fixers has not yet been investigated.

Table 3. Common differences in morphology and symbiotic process of *Frankia* and *Rhizobium* (Diouf *et al*, 2003; Obertello *et al*, 2003)

Feature	*Frankia*	*Rhizobium*
Symbiotic association	Poorly understood	Very well understood
Morphology	• Gram positive • Filamentous • Branching • Slow growing • high G+C DNA content • lacks genetic transformation ability	• Gram negative • Bacillus/rod shaped • Unicellular • both slow and fast growing strains • low G+C DNA content • easily transformed by genetic manipulations
Symbiosis	• Diverse hosts (8 dicot families and 24 genera of plants) • includes woody trees and shrubs • marginal use due to low actual benefit to host plants (max. 288 kg N ha^{-1} in *Casuarina*)	• Single host (members of family Fabaceae) and one nonlegume *Parasponia* • includes herbs, shrubs and trees • 360 kgN ha^{-1} supplied to host plant

ooooo...

□□□□□...

Initial infection events	• undeciphered initial events with marginal information • intracellular root cap invasion or intercellular invasion *via* middle lamella	• well deciphered intricately choreographed initial events • intracellularly by infection thread formation or intercellular invasion *via* spaces between epidermal cells
Nodule	• determinate nodules • Multilobed each representing modified lateral root devoid of root cap • conserve lateral root structure with central VB • formation of prenodule primordia • partially or undifferentiated nodule tissue	• both determinate and indeterminate types • appear stem like due to peripheral VB and central infected tissue • prenodule primordial not formed • nodule tissue well differentiated
Nodule primordia formation	In root pericycle	In root inner cortex (indeterminate type) or outer root cortex (determinate type)
Bacterial cells in nodule	• In peripheral root cortical tissue • two types of cell populations in infected zone	• In central tissue • one type of bacteriod cells in infected zone
Molecular genetics (bacterial)	• No *nod* gene counterparts observed • *nif genes* observed	• Presence of *nod* genes • *nif* genes sequenced
Molecular genetics (host)	• leghaemoglobin protein not formed • ENOD genes not elicited	• leghaemoglobin, forming 13% of nodule weight • rhizobial secretions elicit early and late nodulin (ENOD) genes

Future Perspectives

Actinorhizal plants lure us to know more about the molecular aspects and intricate signaling and interaction biology of *Frankia* with higher woody plant species. Since the isolation and procurement of pure culture of first *Frankia* isolate was possible two decades back several questions remain to be answered regarding this unique symbiosis hampering the identification and development of specific tools for analyzing the molecular biology. Now that the *Frankia* DNA has been cloned in *E. coli* and sequencing is commencing some genes can be identified from a *Frankia* gene library by comparison with other genes. Huguet *et al* (2001) have reported the ability of RFLP-PCR of 16s rDNA gene analysis (phylogenetic analysis) to identify *Frankia* strains from nodules of *Myrica gale, Alnus incana* subsp. *rugosa,* and *Shepherdia canadensis* by comparison of the restriction pattern and sequence of the new isolates with the reference strain. They also concluded that in addition to geographic separation and host selectivity for *Frankia* microsymbionts, edaphic factors such as soil moisture and organic matter content, which varied among locales, may account for differences in *Frankia* populations found in *Myrica* nodules. A major advance in *Frankia*–actinorhizal tree molecular biology will require the use of cloning vectors as well as the development of a transformation system.

References

Akkermans ADL, Hahn D and Baker DD (1992) The family *Frankiaceae. In: The Prokaryotes,* Balows A, Trüper HG, Dworkin M, Harder W and Schleifer KH (eds.), Springer, New York, USA, pp.1069-1084.

Baker D (1987) Relationship among pure cultured strains of *Frankia* based on host specificity. *Physiol Plant,* 70:245-248.

Baker DD and Schwintzer CR (1990) *In: The Biology of Frankia and Actinorhizal Plants* (eds Schwintzer CR and Tjepkema JD), Academic Press, New York, pp.1-13.

Benson DR and Clawson ML (2000) Evolution of the actinorhizal plant symbioses. *In: Prokaryotic Nitrogen Fixation: A Model System for Analysis of Biological Process,* Triplett EW (ed.), Horizon Scientific Press, Wymondham, UK, pp. 207–224.

Benson DR and Silvester WB (1993) Biology of *Frankia* strains, actinomycete symbionts of actinorhizal plants. *Microbiol Rev,* 57:293-319.

Berg RH (1990) Cellulose and xylans in the interface capsule in symbiotic cells of actinorhizae. *Protoplasma,* 159:35-43.

Berry AM and Torrey JG (1979) Isolation and characterization in vivo and in vitro of an actinomyceteous endophyte from *Alnus rubra* Bong. *In: Symbiotic Nitrogen Fixation in the Management of Temperate Forests,* JC Gordon, CT Wheeler and DA Perry (eds.), Corvallis:Oregon State University, pp. 69-83.

Berry MA (1994) Recent developments in the actinorhizal symbioses. *Plant Soil,* 161: 135-145.

Berry MA, Kahn RKS and Booth MC (1989) Identification of indole compounds secreted by *Frankia* HFPArI3 in defined culture medium. *Plant Soil,* 118:205-209.

Bosco M, Fernandez MP, Simonet P, Materassi R and Normand P (1992) Evidence that some *Frankia* sp. strains are able to cross boundaries between *Alnus* and *Eleagnus* host specificity groups. *Appl Environ Microbiol,* 58:1569-1576.

Bulloch BT (1994) Nodulation by *Frankia* increases growth of Casuarinaceae in a New Zealand horticultural soil. New Zealand J Crop Horticultural Sci, 22:39-44.

Clawson ML, Bourret A and Benson DR (2004) Assessing the phylogeny of Frankia-actinorhizal plant nitrogen-fixing root nodule symbioses with *Frankia* 16S rRNA and glutamine synthetase gene sequences. *Molecular Phylogenetics Evolution,* 31:131-138.

Clawson ML, Gawronski J and Benson DR (1999) Dominance of *Frankia* strains in stands of *Alnus incana* subsp. *rugosa* and *Myrica pennsylvanica. Can J Bot,* 77:1203-1207.

Crane PR, Friis EM and Pedersen KR (1995) The origin and early diversification of angiosperms. *Nature,* 374:27-33.

Diouf D, Diop TA and Ndoye I (2003) Actinorhizal, mycorhizal and rhizobial symbioses: how much do we know? *African J Biotecnol,* 2(1):1-7.

Dommergues YR (1997) Contribution of actinorhizal plants to tropical soil productivity and rehabilitation. *Soil Biol Biochem,* 29:931-941.

Duhoux E, Diem HG, Auguy HG, Fernandez D, Bogusz D, Franche C, Dommergues Y and Huguenin B (2001) Angiosperm *Gymnostoma* trees produce root nodules colonized by arbuscular mycorrhizal fungi related to *Glomus. New Phytologist,* 149:115-125.

Gentilli F, Wall LG and Huss-Danell K (2006) Effects of phosphorus and nitrogen on nodulation are seen already at the stage of early cortical cell divisions in *Alnus incana. Annals Bot,* 98(2):309-315; doi:10.1093/aob/mcl109

Gherbi H, Duhoux E, Franche C, Pawlowski K, Nassar A, Berry A and Bogusz D (1997) Cloning of a full-length symbiotic hemoglobin cDNA and *in situ* localization of the corresponding mRNA in *Casuarina glauca* root nodule. *Physiol Plant,* 99:608-616.

Goetting-Minesky MP and Mullin CB (1994) Differential gene expression in an actinorhizal symbiosis: Evidence for a nodule-specific cysteine proteinase. *Proc Natl Acad Sci,* 91:9891-9895.

Guan C (1997) Plant gene expression in actinorhizal nodules of *Alnus glutinosa*. PhD thesis, Wageningen Agricultural University, The Netherlands.

Hahn D, Nickel A and Dawson J (1999) Assessing *Frankia* populations in plants and soil using molecular methods. *FEMS Microbiol Ecol,* 29:215-227.

Han SD and New PB (1998) Variation in nitrogen fixing ability. among natural isolates of Azospirillum. *Microb Ecol,* 36:193-201.

Honerlage W, Hahn D, Zepp K, Zeyer J and Normand P (1994) A hypervariable 23S rRNA region provides a discriminating target for specific characterization of uncultured and cultured *Frankia. Systematic Appl Microbiol,* 17:433-443.

Huguet V, Batzli JM, Zimpfer JF, Normand P, Dawson JO and Fernandez MP (2001) Diversity and specificity of *Frankia* strains in nodules of sympatric *Myrica gale, Alnus incana,* and *Shepherdia Canadensis* determined by *rrs* gene polymorphism. *Appl Environ Microbiol,* 67(5): 2116-2122.

Huss-Danell K (1997) Actinorhizal symbiosis and their N2 fixation. *New Phytol,* 136:375-405.

Huss-Danell K and Myrold DD (1994) Intrageneric variation in nodulation of *Alnus* - consequences for quantifying *Frankia* nodulation units in soil. *Soil Biol Biochem,* 26:525-531.

Laplaze L, Duhoux E, Franche C, Frutz T, Svistoonooff S, Bisseling T, Bogusz D and Pawlowski K (2000) *Casuarina glauca* prenodule cells display the same differenciation as the corresponding nodule cells. *Mol Plant Microbe Interact,* 13:107-112.

Lavire C, Blaha D and Cournoyer B (2004) Selection of unusual actinomycetal primary σ^{70} factors by plant-colonizing *Frankia* strains. *Appl Environ Microbiol,* 70(2):991-998.

Lechevalier MP (1994) Minireview: taxonomy of the genus *Frankia* (*Actinomycetales*). *Int J Syst Bacteriol,* 44:1-8.

Meesters TM, Vliet WM and Akkermans ADL (1987) Nitrogenase is restricted to the vesicles in *Frankia* strain EAN1pec. *Physiol Plant,* 70:267-271.

Normand P, Orso S, Cournoyer B, Jeannin P, Chapelon C, Dawson J, Evtushenko L and Misra AK (1996) Molecular phylogeny of the genus *Frankia* and related genera and emendation of the family *Frankiaceae. Int J Syst Bacteriol,* 46:1-9.

Oakely B, North M, Franklin JF, Hedlund BP and Staley JT (2004) Diversity and distribution of *Frankia* strains symbiotic with *Ceanothus* in California. *Appl Environ Microbiol,* 70(11): 6444-6452.

Obertello M, Sy MO, Laplaze L, Santi C, Svistoonoff S, Auguay F, Bogusz D and Franche C (2003) Actinorhizal nitrogen fixing nodules: infection process, molecular biology and genomics. *African J Biotechnol,* 2(12):528-538.

Pawlowski K, Twigg P, Dobritsa S, Guan CH and Mullin B (1997) A nodule specific gene family from *Alnus glutinosa* encodes glycine- and histidine-rich proteins expressed in the early stages of actinorhizal nodule development. *Mol Plant Microbe Interact,* 10: 656-664.

Quispel A, Svendsen AB, Schripsema J, Baas WJ, Erkelens C and Lugtenburg J (1989) Identification of a dipterocarpol as isolation factor for the induction of primary isolation of *Frankia* from root nodules of *Alnus glutinosa* (L.) Gaertner. *Mol Plant Microbe Interaction,* 2:107-112.

Richards JW, Krumholz GD, Chval MS and Tisa LS (2002) Heavy metal resistance patterns of *Frankia* strains. *Appl Environ Microbiol,* 68(2):923-927.

Selim S, Delacour S and Schwencke J (1996) Specific long-chain fatty acids promote optimal growth of *Frankia*: accumulation and intracellular distribution of palmitic and propionic acid. *Arch Microbiol,*165: 252-257.

Selim S, Delacour S and Schwencke J (1996) Specific long-chain fatty acids promote optimal growth of *Frankia*: accumulation and intracellular distribution of palmitic and propionic acid. *Archives Microbiol,* 165, 252-257.

Sims HJ, Herendeen PS, Lupia R, Christopher RA, Crane PR (1999) Fossil flowers with Normapolles pollen from the Upper Cretaceous of southeastern North America. *Rev Palaeobot Palyn,* 106:131-151.

Smith SE and Smith FA (1990) Structure and function of the interfaces in biotrophic symbioses as they relate to nutrient transport. *New Phytologist,* 114:1-38

Tarrant RF and Trappe JM (1971) The role of *Alnus* in improving the forest environment. *Plant Soil,* Special Vol: 335-348.

Tisa LS, Chval MS, Krumholz GD and Richards J (1999) Antibiotic resistance patterns of *Frankia* strains. *Can J Bot,* 77(9):1257-1260.

Tunlid A, Schultz NA, Benson DR, Steele DB and White DC (1989) Differences in fatty acid composition between vegetative cells and N_2-fixing vesicles of *Frankia* sp. strain Cpu1. *Proc Natl Acad Sci USA,* 86:3399-3403.

Verghese S and Misra AK (2002) *Frankia*–actinorhizal symbiosis with special reference to host–microsymbiont relationship. *Curr Sci,* 83(4): 404-408

Wall LG (2000) The actinorhizal symbiosis. *J Plant Growth Regul,* 19:167-182.

Wheeler CT, Tilak M, Scirmgeour CM, Hooker JE and Handley LL (2000) Effects of symbiosis of *Frankia* with arbuscular mycorrhizal fungus on the natural abundance of ^{15}N in four species of *Casuarina. J Exp Botany,* 59(343):287-297.

Yamanaka T, Li CY, Bormann BT and Okabel H (2002) Tripartite associations in an alder: effects of *Frankia* and *Alpova diplophloeus* on the growth, nitrogen fixation and mineral acquisition of *Alnus tenuifolia. Plant Soil,* pp.1-8 (Author's proof) cited online. M:Disk CP PIPS NO.:5111671 (plsokap:bio2fam) v.1.1.

Yang W, Horvath B, Hontelez J, van Kammen A and Blsseling T (1991) *In situ* localization of *Rhizobium* mRNAs in pea root nodules: *nifA* and *nifH* localization. *Mol Plant-Microbe Interact,* 7:276-281.

Zepp K, Hahn D and Zeyer J (1997a) Evaluation of a 23S rRNA insertion as target for the analysis of uncultured *Frankia* populations in root nodules of alders by whole cell hybridization. System *Appl Microbiol,* 20:124-132.

Zepp K, Hahn D and Zeyer J (1997b) *In situ* analysis of introduced and indigenous *Frankia* populations in soil and root nodules obtained on *Alnus glutinosa. Soil Biol Biochem,* 29:1595-1600.

Mind Bogglers

1. What are the plants, which are nodulated by actinomycetes to cope up poor or soil deficient in nitrogen nutrition?
2. Enlist the major characters used for classification of *Frankia*?
3. How the nodule forming process in actirhizal roots differs from the nodulation sequence or steps in rhizobia? Also quote the similarities between these two symbioses.
4. Briefly explain the various characters and processes differing in rhizobia and *Frankia*.
5. Give a detailed account of various molecular techniques for detection and characterization of various frankial microsymbionts.

3

Cyanobacterial Symbiosis with Vascular Plants

- Introduction 62
- Cyanobacterial Systematics 64
- Plant Host for Cyanobacterial Symbiosis 65
- Molecular Biology of Cyanobacterial Symbiosis 70
- Role as Biofertilizer 73
- Mass Production Technology 75
- Future Perspectives 78
- References 81
- Mind Bogglers 84

Introduction

Nitrogen-fixing cyanobacteria are the group of most primitive prokaryotes that have been observed in proterozoic era. Many Proterozoic oil deposits are attributed to the activity of Cyanobacteria, such as *Gloeocapsomorpha*. Small concentrically layered structures called pisolites are also the result of fossilized bacteria.

- Palaeontological occurrence of prokaryotes 3,450 million year ago
- Cyanobacteria among pioneer organisms of early earth
- First autotrophs as primary organic matter producers that released elemental oxygen into reducing primitive environment
- Very versatile organisms that inhabit variety of diverse environmental niches as
  - **aquatic** (brackish/freshwater, limnic/marine, saline sodic estuaries, euryhaline aquifers),
  - **land** (soil, bare rocks),
  - **infertile/hostile extreme environmental niches** hot (thermal vents) springs, mountain streams or arctic/Antarctic ice glaciers, desert sand dunes, rocks, volcanic ash
  - **distribution** plankton/pelagic/benthic (marine environment),
  - **diverse trophic states** in fresh water habitats near-surface epilimnic and deep, euphotic, hypolimnic waters of lakes or surfaces by attaching to rocks or sediments, forming mats floating on surface.
  - **various interaction state** as free living to a range of symbiotic interactions with diverse genera of lower and higher plants Schopf (1996)

These group of bacteria enter into symbiotic associations with phylogenetically diverse groups of eukaryotes ranging from algae, fungi, bryophytes (hornworts and liverworts), pteriodophytes (ferns *Azolla*) to higher vascular plants as gymnosperms (Cycads) and angiosperms (*Gunnera*) (Rai *et al*, 2000) which are stable over a wide range of environmental conditions. Cyanobacteria show symbiosis with lower forms of plants. The fresh water diatom *Rhopalodia gibba* is unique in showing symbiosis with a unicellular cyanobiont located intracellularly in the photosynthetic host that fixes nitrogen (Drum and Pankratz, 1965). The cyanobacterial symbionts have been reported to colonize feather moss *Pleurozium schreberi* and morphological and molecular identification studies of isolates showed presence of three different cyanobacteria *Nostoc* sp., *Calothrix* sp. and *Stigonema* sp. with the last one to be predominant moss colonizing isolate (Gentili *et al,* 2005). Majority of cyanobacterial symbiotic interactions are facultative and the bacterial cells remain extracellular in macrosymbiont plant host while the exceptions as *Gunnera* let the cyanobacteria to reside intracellularly in specialized organs formed during early events of symbiosis. Cyanobacterial species exhibiting symbiosis share a wide biotrophic exchange of metabolites with host plant as partners leading to harmonious proliferation of the both symbionts as a single stable unit. These autotrophic diazotrophs play a major role in furnishing reduced atmospheric nitrogen to the host plant in form of NH_4^+ ions for enhanced protein synthesis and biomass production.

Symbiotically competent cyanobacteria have some excellent features that make them particularly significant in any attempt to extend the list of N_2-fixing symbioses to include plants of commercial interest, such as cereals. Unlike rhizobia (the symbionts of legume plants), most symbiotic cyanobacteria carry heterocysts (heterotrophic specialized cells) to combat the nitrogenase inactivation by oxygen. These symbionts communicate using chemical signals like rhizobia and require elicitation of both the genes of micro as well as macrosymbiont during the establishment of a functional N_2-fixing symbiosis. The elicitation of genes and formation of new proteins and other compounds that would largely assist or result in establishment of symbiosis occur in both partners. In general *Nostoc* is the most common symbiotic cyanobacterial genera but the others include *Anabaena, Calothrix* and *Chlorogloeopsis*.

Cyanobacterial Systematics

Cyanobacteria have been classified and named both under Botanical (International code of botanical nomenclature, ICBN) as well as Bacteriological (International code for nomenclature of bacteria, ICNB) coding systems and thirteen species (till May 2003) have been reported and published in International Journal of Systematic Bacteriology (Oren, 2004). Blue green algae or cyanobacteria are microscopic, prokaryotic cells with autotrophic mode of nutrition, lack flagella, exhibit presence of phycobilin pigments (C-phycocyanin and C-phycoerythrin), store food as myxophycean starch or cyanophycin, contain lamellar structures and heterocysts and completely lack sexual reproduction as reproduce vegetatively by fission, fragmentation or by formation of hormogones. These life forms have an extraordinarily primitive lineage and ancestors of cyanobacteria date back in Precambrian archaeozoic era. All members belong to single class cyanophyceae consisting of five orders, fourteen families, 150 genera and approximately 2,500 species.

Prokaryotes as cyanophytes being morphologically simplified forms have been classified previously largely based on physiological properties (culture characteristics) in pure laboratory cultures (Komarek, 2003) while morphological properties overweigh under field studies that include cell shape, dimensional differences, cell division patterns, colony formation and relationship to extracellular envelopes and sheaths. Broadly cyanobacteria on basis of morphology could be classified as heterocystous cyanobacteria and non-heterocystous or unicellular cyanobacteria. Among the former group there occurs spatial separation between oxygenic photosynthesis which is carried out in vegetative cells and nitrogen fixation that is performed by virtue of specialized non-photosynthetic heterocysts formed during nitrogen starvation conditions. The nitrogenase in heterocysts remain protected from oxygen inactivation owing to increased respiratory rate or decreased oxygen permeability. While non heterocystous ones exhibit temporal separation of the two processes (photosynthesis in day and nitrogen fixation in night) and thus have not very well defined mechanisms of oxygen control during nitrogenase functioning.

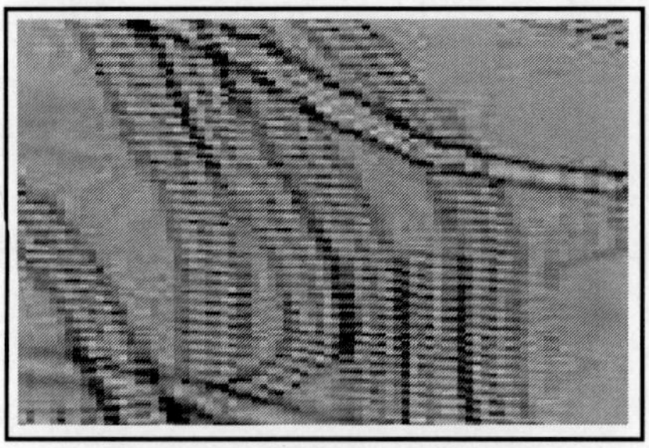

Spirullina cells under light microscope

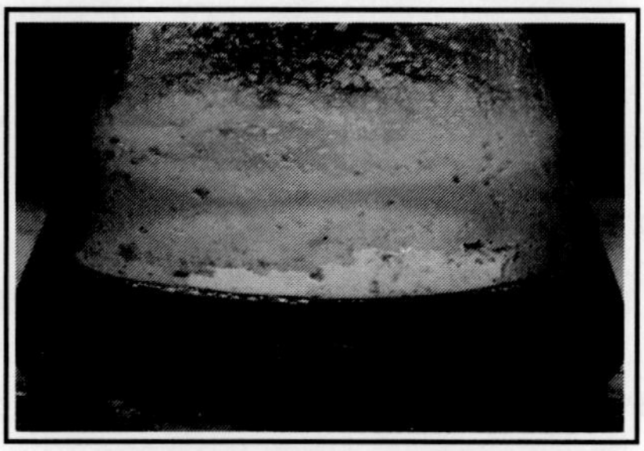

Nostoc culture in liquid Fogg's medium

(For a coloured version of this figure, see plate section, page 299)

Though useful in classifying cyanobacteria at field level this was rather inefficient system as new forms/strains develop as well as there occurs considerable variability in morphological features with in different environmental conditions (Oberholster *et al,* 2005). New technological advancements have furnished novel criteria to further classify cyanobacteria at DNA/RNA levels. Cyanobacterial taxonomy has evolved over the decades and has undergone considerable changes owing to input from several branches of taxonomical analyses as morphology, cytology, cultural characters, physiology and biochemistry and recently the PCR based techniques particularly rRNA phylogenetic criterion.

Plant Host for Cyanobacterial Symbiosis

Cyanobacteria exist in symbiotic association with an array of forms ranging from fungal lichens as *Peltigera* (O'Brien *et al,* 2005), eukaryotic algae (Murakami *et al,* 2004), bryophytes (Gentili *et al,* 2005), pteridophytes, gymnosperms to angiosperms (Rai *et al,* 2000) and animal forms as sponges (Steindler *et al,* 2005) which foretells the versatility of these prokaryotic autotrophic forms. The various plant hosts that undergo cyanobacterial symbiosis are discussed below.

i) *Bryophytes:* Cyanobacteria exhibit symbiosis with 4-6 genera of hornworts (*Phaeoceros*) and 2 genera of liverworts (*Blasia*) and reside extracellularly in specialized cavities of gametophytes where the actual nitrogen fixation is carried out in term of spatial differentiation of vegetative and heterocystous cells (frequency 25-45%). Lower surface of *Anthocerous* thallus harbours *Nostoc* that stimulates papillae formation providing space for formation of compact colonies of alga within the cavities of host. Liverwort genera *Blasia* and *Cavicularia* exhibit symbiosis with *Nostoc sphaericum* that inhabit cavities on outgrowths near edges of lower surface. Cyanobacterial species *Nostoc* and *Calothrix* were also isolated and identified both microscopically and molecularly by Gentilli *et al* (2005) from feather moss *Pleurozium schreberi* samples collected from northern Sweden. Villarreal and Renzaglia (2006) have reported a novel symbiotic arrangement in land plants i.e. occurrence of *Nostoc* cells in longitudinally oriented schizogenous canals of *Leosporoceros dussii*, member Anthoceratophyta. In surface view, the microsymbiont cells (*Nostoc* canals/strands) appear as elongated, dichotomously branched blue-green strands ocuring at the center of the thallus.

ii) *Pteridophytes:* One genera among pteridophytes *Azolla* exhibits most unique cyanobacterial symbiosis with the genus *Anabaena* residing extracellularly in the specialized cavities of dorsal leaves with a very low heterocyst frequency of 0-30% (Meeks and Elhai, 2002). *Azolla* leaf consists of a thick, greenish (or reddish) dorsal (upper) lobe and a thinner, translucent ventral (lower) lobe that remain immersed in the water. The ovoid central cavity of the upper lobe is inhabited by *Anabaena* filaments. This endophytic association allows nitrogen fixation when *Anabaena* resides in the ventral pore of *Azolla* fern.

The *Azolla* cyanobiont has been isolated and identified however the prior reports predict a general trend of difference in antigenic and molecular genetic properties (by DNA fingerprinting) of cultured *A. azollae* from freshly isolated endosymbionts. Even inoculation of the cultured *Anabaena*

Order	Family	Representative genera	Unique characters	Reference
Chroococcales	Chroococcaceae Entophysalidaceae	*Chroococcus* *Entophydalis*	• Unicellular or colonial forms without trichomes • reproduction by binary fission	Rippka *et al* (1979)
Chaemaesiphonales	Chaemaesiphonaceae	*Chamaesiphon*	• Unicellular or colonial forms without trichomes • reproduction by budding or multiple fission	Fritsch (1945)
Pleurocapsales	Pleurocapsaceae	*Pleurocapsa* *Hyella* *Myxosarcina*	• filamentous • no trichome formation • heterocysts absent • reproduction by endospore formation	Fritsch (1945)
Nostocales	Oscillatoriaceae Nostocaceae Syctonemataceae Microchaetaceae Rivulariaceae	*Oscillatoria* *Anabaena* *Scytonema* *Fortica* *Calothrix, Rivularia*	• filamentous, • trichome present • pseudobranching common • reproduction by trichome fragmentation or by hormogonia or akinetes	Rippka *et al* (1979)
Stigonematales	Capsospiraceae Nostochopsidaceae Mastigocladopsidacea Mastigocladaceae Stigonemataceae	*Capsospira* *Mastigocolus* *Mastigocladopsis* *Mastigocladus* *Stigonema*	• filamentous • true branching • heterocystous forms • reproduction by trichome fragmentation, hormogonia and rarely akinetes	Rippka *et al* (1979)

strains into germinating sporophytes of *Anabaena*-free *Azolla* sp. failed to establish N_2-fixing symbiosis (Lin *et al,* 1989). However the later work by Tang *et al* (1990) of limited multiplication of this cyanobiont in artificial media containing fructose and Casamino acids under 1% oxygen conditions established that the isolated forms were antigenically similar to the endosymbiotic forms. This work was further carried on by Pabby *et al* (2004) by performing a comparative study of the morphological attributes of freshly separated and cultured cyanobionts of six species of *Azolla,* namely *A. microphylla, A. filiculoides, A. rubra, A. caroliniana, A. mexicana* and *A. pinnata.* They reported distinct variation with respect to color and nature of filaments, size and frequency of heterocysts of cultured and freshly separated cyanobionts from each species of *Azolla.*

Indian *Azolla pinnata* is used as common green compost and produces higher biomass (200-300 t ha/yr) than the popular green manures as *Crotolaria, Tephrosia* or *Sesbania* (30-50 t ha/yr). The *Azolla-Anabaena* symbiosis proves to be congenial biofertilizer for rice based cropping systems being the major components of wetland rice ecosystems, which are easily available and serve as the cheapest sources of natural biofertilizers.

Limitations: Azolla addition as organic manure for rice cultivation is useful and mass cultivation of *Azolla* requires extensive water application, which is a limitation for water scarce areas. The process is labour intensive and cost effective in only developing countries where labour deputation is low cost. Moreover high amount of biomass production remains low due to the problem of autolysis as well as biology or factors promoting spore formation are less understood. The spore production could help in overruling these limitations. This plant is also susceptible to parasites and sensitive to fluctuations in temperature thus posing a problem for its wide spread use in the northern states as Punjab where summer (kharif season) temperatures hike upto 45-47°C and fluctuate throughout the cropping season. Above all the phosphorus applied to the rice plants as single super phosphate (SSP) causes stunting of the *Azolla* plants as well as decreased/nil nitrogenase activity and disappearance of *Anabaena azollae* from *Azolla* fronds due to toxic effect of applied P fertilizer to the microsymbiont (Dawar and Singh, 2002).

iii) *Gymnosperms (Cycads):* These include 11 genera of cycas that have the affinity to form symbiotic associations with cyanobacteria particularly heterocystous members of order Nostocales *viz. Nostoc* or *Anabaena,* extracellularly in a layer in specialized coralloid root structures (Lindblad and Costa, 2002). This association is symbiotic as the host provides the photosynthate to the cyanobacteria resident in corraloid roots while the microsymbiont generously fixes and assimilates nitrogen to the host. This association could be mimicked/ reconstituted in lab conditions as done by Owe *et al* (1999) for *Zamia furfuracea.*

Cycas plant exhibits symbiosis with BGA

Nostoc and *Anabaena* in Fogg's medium

(For a coloured version of this figure, see plate section, page 299)

iv) *Angiosperms (Gunnera):* It is a herbaceous flowering plant (Anthophyta) comprising 40 species known to host filamentous motile forms of cyanobacteria (Hormonogonia) specifically *Nostoc* intracellularly in specialized stem glands of the plant (Chiu *et al,* 2005). It exhibits occurrence of characteristic mucilage filled cavities or glands near petiole base. The specialized glands (red due to accumulation of anthocyanins as pelargonidin and cyanidin) on the *Gunnera* stem serve as entry points for *Nostoc* filaments. Subsequently, one new gland typically appears on the stem at the base of each new leaf. Under natural conditions, all of the glands are colonized by cyanobacteria (Osborne *et al*, 1992). *Nostoc* filaments remain intracellular surrounded with a membrane probably of host plasma membrane origin and the microsymbiont provides reduced ammonia to host.

Heterocyst: Specialized rectangular or nearly round cells having thickened non cellulosic polysaccharide cell wall, homogenous transparent pale yellowish cytoplasm present at regular intervals (terminal or intercalary positions) in filamentous trichome forms

Hormogonia: Specialized motile cells formed by breakage of trichomes of filamentous cyanobacterial forms within sheath

Akinetes: Specialized large thick walled yellow to brown coloured cells of several shapes (spherical, discoid, elongated) showing resistance to dessication and meant for perenation of heterocystous forms under unfavourable conditions

Molecular Biology of Cyanobacterial Symbiosis

The genome size of representative cyanobacteria ranges from 1.6 to $8.6x10^9$ Da compared to that of other bacteria (1.0 to $3.6x10^9$ Da) (Herdman *et al*, 1979). Cyanobacterial genome analysis has increased our understanding of biochemical and physiological aspects of nitrogen fixation by these bacteria however paradoxically a little is known regarding genetics of nitrogenase enzyme. The cyanobase (databank including the nucleotide sequence of the whole genome of cyanobacteria) could be helpful in comparing and elucidating the base sequence of the cyanobacterial species sequenced till date (eight genera). Kaneko *et al* (1996) have performed the whole genome sequence of unicellular cyanobacterium *Synechocystis* sp. strain PCC6803 so that general genetic complement and the specified genetic information necessary for oxygenic photosynthesis in this alga could be obtained. They reported the genome size to be 3.6 Mb after construction of a fine physical map of *Synechocystis* genome. A similar report by Nakamura *et al* (2002) provided genetic information of *Thermosynechococcus elongatus* BP-1 strain obtained from marine environment. These reports would help in establishing a correlation of these photooxic bacteria with the other diazotrophic counterparts be it free living, associative or symbiotic nitrogen fixers.

Recent evidence suggest that some heterocystous cyanobacteria contain two molybdenum-based nitrogenase systems, *nif1* and *nif2*, with different expression patterns. The vegetative cell nitrogenase (*nif2*) has its counterpart in non-heterocystous cyanobacteria that fixes N_2 only anaerobically. The nitrogen fixation (*nif*) genes in vegetative cells of heterocystous cyanobacteria, such as *Anabaena,* exist as two operons that have two separate intervening DNA elements *viz* an 11kb element (GGATTACTCCG) directly repeated at ends disrupts *nifD* gene in the *nifHDK* operon that harbours *xisA* (codes for a site specific recombinase) and a large 55 kb element containing *xisF* gene (codes for another site specific recombinase) and two 5 bp direct repeats (TATTC) at ends interrupts the *fdxN* gene in the *nifBfdxNnifSnifU* operon resulting in the dispersion of genes and impaired gene expression. During heterocyst differentiation both the discontinuities are precisely excised by two distinct site specific recombination events. One of them is brought about by the XisA protein between the 11 bp direct repeats. The second one is caused by the XisF protein and occurs between the 5 bp direct repeats. As a consequence the 11kb and 55 kb elements are removed from the chromosome as circles and functional *nif* operons are created (Apte and Prabhavathi, 1994). Furthermore, some cyanobacteria may produce Mo-independent 'alternative' (V or Fe) nitrogenases, encoded by the *vnf* and *anf* genes. Nitrogenase proteins are then expressed from the rearranged genes in heterocysts and aerobic nitrogen fixation ensues.

The genetic regulation of nitrogen fixation in cyanobacteria differs from that in other N_2-fixers. The presence of a source of combined nitrogen like nitrate or ammonium in the environment generally prevents cyanobacteria from fixing nitrogen as these microbes fail to synthesize nitrogenase in presence of nitrate or ammonium. An NtcA transcriptional regulator regulates the expression of nitrogen assimilation genes such as *het*R and *gln*A genes that are involved in heterocyst differentiation and ammonia assimilation. Moreover another novel protein PII protein (encoded by *glnB*) has been characterized that may have probable function of a signal transducer/sensor during N stress conditions and thus help in coordination of C and N-assimilation. Another novel observation regarding cyanobacterial nitrogen fixation is regarding ability of several cyanobacteria particularly unicellular *Gloeothecae* to fix nitrogen in intermittent light conditions that helps in assimilation of the nitrogen fixed under dark conditions and thereby avoidance of nitrogenase inhibition by accumulated nitrogen.

Expanding our understanding of the genetics of free-living and symbiotic organisms (e.g. *Gloeothecae, Trichodesmium, Nostoc and Calothrix*) would help in development of an appropriate genetic methodology in easy manipulation of these cyanobacteria. There are new reports coming on the symbiotic interaction and exchange of compounds among symbiotic cyanobacteria with their respective hosts though our understanding and knowledge about the mechanisms involved in exchange of nitrogen fixed by cyanobacteria to host plant are scarce. Meeks (1998) has reported the production of certain elicitor molecules by the plant partner that control the differentiation of hormogonia and heterocysts. A similar report by Meeks *et al* (2002) hypothesized that heterocyst, akinete and hormogonium differentiation in *N. punctiforme* and *Anabaena* sp. strain PCC 7120 involve complex developmental regulatory cascades similar to *Bacillus* or *Myxococcus* sporulation. The pivotal elements of regulatory signal cascades could be partly identified by comparative bioinformatics analysis of the sequenced genomes of organisms and partly by functional analysis.

A decrease in the rates of three physiological processes i.e. growth, assimilation of CO_2, and assimilation of NH_4^+ have been documented on entrance of cyanobacteria in symbiotic state while an increase occurs in the rate of N_2 fixation. Physiological behaviour in terms of regulation of size of symbiotic *Nostoc* colony and nitrogenase activity occurs in *Anthoceros*. A similar reaction occurs in the microsymbiont cyanobacteria i.e. decrease in photosynthesis and reduction of CO_2 to organic carbon (Meeks, 1998). Similarly the rate of light-dependent CO_2 fixation remains generally low by freshly separated symbiotic *Nostoc* than in free-living cultures except *Anabaena azollae*.

Role as Biofertilizer

Cyanobacteria have a marked role as biofertilizer by virtue of an array of mechanisms which primarily include biological nitrogen fixation that provides atmospheric nitrogen in reduced NH_3 form to plant as well as by production and secretion of plant growth promoting substances. The reduced N (as NH_3) fixed by algae is released either through exudation or through microbial decomposition after cell death. This fixed N helps in increasing the growth and yield of the inoculated plants and may remain in the soil as accumulated N pool by N buildup or may be used by the plant. On an average the plant body may contain approximately 25-30% of nitrogen fixed by cyanobacteria as reported by Wilson (1980) who recovered 39% of the nitrogen from ^{15}N-labelled *Aulosira* spp. spread on the soil from rice crop and 51% from the algae incorporated into the soil. The benefit of *Azolla* addition to rice fields at a rate of 10 t/ha is equivalent to 25 to 30 kg N/ha (Mishra and Pabbi, 2004). Carrapico *et al* (2000) reported 50% savings of the chemical fertilizer on application of *Azolla* compost to the rice fields in Africa without significant loss in rice yield. However a range of abiotic (chemical like pH, Eh, organic matter, mineral composition and climatic like temperature, soil moisture) and biotic factors (parasitism by microarthropod predators, antagonism with native microflora) affect the relative percentage of the total nitrogen fixed by BGA in paddy fields which remains variable enough.

Cyanobacteria can be incorporated into soil as organic matter and also as a source of enzymes as they produce acid and alkaline extracellular phosphatases that are active in solution or located in the periplasmatic space of the cell wall. Both biomass and exopolysaccharides incorporated into soil, induce growth promotion of other microorganisms and increases the activity of soil enzymes that participate in the liberation of nutrients required by plants (Caire *et al*, 2000).

In developing countries as India algalization of rice fields has a positive effect for the new high nitrogen responsive rice varieties even in presence of high levels of fertilizer nitrogen which could be attributed to growth promoting substances (GPS) synthesized by these bacteria that are readily absorbed and utilized by the crop plants. The common biologically potent substances secreted by cyanobacteria include essential amino acids, proteins, and sugars, vitamins (Vit B_{12}), phytohormones as IAA that are utilized to an extent by the bacteria for its growth while the remaining are released out of the cells in the surroundings as excretions during growth cycle or after cell death and contribute towards enhanced plant growth. Haroun and Hussein (2003) have reported a significant increase in auxin, gibberellic acid and cytokinin phytohormones while a significant decrease in abscisic acid content of *Lupinus termis* plants grown in siliceous soil.

Apart from these benefits the soil physical and chemical properties are altered by algalization owing to addition of algal biomass that produces polysaccharides to improve soil aggregation. While the soil chemical properties are also changed due to enhanced ammonification that increases the available nitrogen, enhanced soil organic carbon (2-11%) and solubilization of unavailable P reserves due to secretion of organic acids to make available phosphorus to plants. Algalization also reduces the amount of exchangeable sodium and thus alters the pH of the soil after application and thereby helps in reclamation of saline or alkali/sodic soils and improves soil fertility.

Various plant growth promoting substance synthesized by cyanobacteria

PGPS produced	Genera
Gibberellin	*Calothrix* sp, *Anabaena* sp, *Stratonostoc* sp, *Aulosira* sp
Indole-3-acetic acid or other auxins	*Anabaena* sp, *Nostoc* sp, *Cylindrospermum muscicola*
Vitamin B_{12}	*Cylindrospermum muscicola*
Ascorbic acid	*Nostoc* sp., *Anabaena* sp.
Pantothenic acid, niacin, folic acid	*L. plantarum, L. leichmanii, S, faecalis*
Amino acids (cysteine, tyrosine, phenylalanine)	*Anabaena* sp, *Nostoc* sp,

Mass Production Technology

Owing to increasing costs of the chemical fertilizers and low fertilizer use efficiency and increasing tendency for sustainable crop production the development of these algal biofertilizers for enhanced crop yield particularly rice production has increased over the years. However development of biofertilizer technology has impeded due to lack of information on sustenance of inoculated strains by climatic and biotic factors in rice fields and also due to inability to efficiently produce good quality inocula. The inoculation time also affects the total benefit that could be harnessed by algalization. Healthy inoculum, inoculated by either broadcasting of powdered inoculum, sand (prewashed) mixed inoculum or by application of fresh inoculum or liquid inoculum, on shiny sunny day would help in better development and multiplication of alga than during cloudy days.

a) Open pit/pond/tray/trough technique: The basic method of mass production involves a mixture of nitrogen fixing cyanobacteria in shallow trays or polythene lined pits filled with water kept in open air, using clean, sieved farm soil as a carrier material. After addition of single super phosphate (SSP) and water starter culture, a mixture of *Anabaena, Nostoc, Aulosira* and insecticides (carbofuran, malathion or BHC) are added to prevent insect breeding. Insecticides are added for common predators of cyanobacteria that include the chytridious fungi, myxobacteria, cyanophages, snails, copepods (*Daphnia*) and insect larvae particularly the mosquito larvae. The cyanobacterial growth is observed after 10-12 days as thick mat, which is collected, dried and packed as inocula to be added to the rice fields or the dried mass in open field culture technique is left for few days and the flakes so obtained could be applied at a rate of 10 kg/ha as inoculum in the rice fields.

Constraints

- culture is prone to be contaminated by green and diatomous algae
- entire process is labour intensive
- cultures had to be kept at relatively low cell density to obtain high yield
- require frequent monitoring regarding the establishment of the BGA culture
- biomass application after harvesting from pond requires processing, which reduces, cell viability and culminate to increased cost of production

b) *Commercial straw based production technique:* The commercial straw based production could be performed in poly or glass houses under ambient temperature (25-30°C), light intensity and humidity in polythene lined shallow pits as soil based cultivation of BGA. The starter culture is added followed by multiplication for around a week. The produce is harvested, mixed with rice or wheat straw as carrier, sun dried and packed in polybags.

c) *New techniques for mass production:* To overrule the problems of open tray/field mass production cyanobacteria could be cultivated at high cell density in reactor that ensures high yield.

i) *Closed tubular reactor:* This reactor could be utilized to obtain significantly higher rates of biomass production having higher edge over biomass produced by traditional open pond cultivation technique by overcoming the problem of sub optimum temperatures that prevail during later technique. The blue green algal mass thus produced in closed tubular reactor requires no processing protocols to be followed and could be applied directly to fields as dense cultures thus reducing the cost of production.

Advantages

- It ensures increased rate of survival of formed algal mass as drying is not followed by production
- It overrules the problem of culture dilution as in case of open pond method where there remain chances of wash-off of cultures. Hence closed reactor ensures maintenance of concentration of mass developing even during rainy season
- Large scale and prolonged algal biomass production help in use of produce as animal, fish or poultry feed

ii) *Carrier based immobilized inoculant production:* The most popular technique of cyanobacterial immobilization includes immobilizing *Anabaena* sp cells on polyurethane foam however the other alternatives have also been utilized that included the sugarcane waste or bagasse and paper waste that are mixed with rice husk and soil mixture as carrier to form ready to use bioinoculant with enhanced ability of ammonia release and enhanced crop chlorophyll, total carbohydrate and protein. Thus this has a better edge over traditional BGA inoculant cultures that do not exhibit enhanced photoproduction of ammonia under flooded conditions in low land rice cultivation protocols.

Advantages

- Immobilization avoids readily wash off during frequent rains
- Continuous availability of ammonium ions
- Enhanced PGP action due to steady and continuous release of PGPS from immobilized culture

FUTURE PERSPECTIVES

Cyanobacteria are unique group of bacteria that fix nitrogen under nitrogen deprivation and new insights provided by sequence analysis of *Anabaena* and *Nostoc* have enlightened our sketchy knowledge regarding elaborate control mechanisms of nitrogenase gene translation as well as processing and control of activity at the enzyme level. Still more accurate reports of *in situ* nitrogen fixation rates would clear our version of nitrogen fixation efficiency and utility of cyanobacteria as a bioinoculant in sustainable production of cereal crops particularly nitrogen input intensive paddy crop. The genetic analysis is also useful in marking the presence of specific enzymes required in production of plant growth promoting substances (PGPS) during growth of BGA (Thajuddin and Subramanian, 2005). The DNA sequence information would further help in easier manipulation of nitrogen fixation and PGPS genes for formation of newer strains with enhanced nitrogen fixation capabilities as well as better PGP action. The research for production and development of new region specific strains as well as potent strains with higher establishment would be an asset for the new age farmers. Moreover research is prerequisite in field of production of inorganic nitrogen fertilizer or chemical nitrogen application strategies that support or exhibit little impact on biological nitrogen fixation by BGA.

The stress tolerance of cyanobacteria has always been the weak point for their use in harsh tropical/subtropical conditions where temperature hikes are observed. Several reports are being published on the role of lipids in conferring tolerance to several physiological stressors in cyanobacteria and the whole mechanism of desiccation tolerance in cyanobacteria relies on phospholipids bilayers, which are stabilized during water stress by sugars, especially by trehalose. Trehalose sugar prevents damage from dehydration not only by inhibiting fusion between adjacent membrane vesicles during drying, but also by maintaining membrane lipids in the fluid phase in the absence of water (Singh *et al*, 2002). These molecule level studies of the cell wall organizational chemistry would help in developing new more dessication resistant strains of cyanobacteria.

- Cyanobacteria also possess bioactive compounds of medicinal importance
- Two types of phospholipids *viz* acetylsphingomyelin and an acylacetylglycerol phosphoacetylated glycolipid were isolated from *Scytonema julianum* causing platelet aggregation
- A new type of lipid in cyanobacteria exhibiting biological activity very similar to that of PAF has been isolated (Antonopoulou *et al,* 2002)

The other facet of science exhibiting information pour in is the molecular biological and genomics. The cyanobacterial genome has been sequenced by shotgun technique (*http://www.jgi.doe.gov*) and a preliminary analysis has been published. The genome analysis has further led to the identification of signals in the developmental biology of cyanobacteria. Cyanobacterial sequence analysis has resolved the much talked about problem of origin of chloroplasts in eukaryotes. A fine review by Raven and Allen (2003) has provided firm support to the theory of endosymbiosis by cyanobacteria as the cause of origin of chloroplasts in eukaryotes.

The cyanobacterial biodiversity analysis is springing up as a promising branch of microbial diversity discipline. Molecular marker analysis is becoming increasingly capable of identifying informative genetic variation. A study by Oberholster *et al* (2005) has revealed the role of amplified fragment length polymorphism markers (AFLPs) to differentiate between geographically unrelated *Microcystis* strains and its use/applicability in cyanobacterial taxonomy particularly providing abundant information on the phylogenetic status and potential in evolutionary studies. Steindler *et al* (2005) have reported polyphyletic nature of origin of sponge associated cyanobacteria (derived from multiple independent symbiotic events) after phylogenetic analysis of the 16s rRNA sequences.

Several reports regarding the Indian cyanobacterial diversity of various niches as top soil crust, marine/freshwater environment and extreme conditions suggest that our environmental conditions nurture a wide range of cyanobacteria (Tirkey and Adhikary, 2005; Thajuddin and Subramanian, 2005). The microbial diversity analyses led to advancements in tools and techniques for deciphering novel as well as previously characterized forms in rapid and automated fashion. Cyanobacterial identification criteria have also evolved over the years and are prerequisite owing to their power of deciphering the presence of indigenous or the marked establishment of the inoculated strains. Various DNA based/ molecular marker analysis protocols including the AFLP, PCR-RFLP and rRNA oligotyping approaches are able to trace the presence or absence of these bacteria in a given niche or sample using specific conserved DNA sequences (Worden *et al,* 2000). Sequence analysis of the DNA-dependent RNA polymerase (*rpoC1*) gene from 24 *Anabaena* isolates, including 12 designated *A. circinalis,* permitted a phylogenetic analysis to be performed. In addition, an *A. circinalis*-specific PCR was developed and tested successfully on environmental samples (Fergusson and Saints, 2000). Guevara *et al* (2002) have performed PCR fingerprinting of short tandemly repeated repetitive sequences of *Nostoc* infected symbiotic tissue of *Gunnera tinctoria* and reported genetic diversity of symbiotic *Nostoc* among host plants as well as observed strain specificity phenomena for this particular symbiosis (i.e. one or closely related strains undergo symbiosis with an individual host plant).

The cyanobacterial association has also been tried to extend to non-legumes (cereals) so as to equip these plants (rice, wheat) to obtain some part of their N nutrition by associated cyanobacteria by BNF (Rai and Bergman, 2002). Coculture of *Nostoc* or *Anabaena* strains with wheat seedlings causes root colonization and seedling growth (Rai *et al,* 2000).

References

Antonopoulou S, Olkonomou A, Krantonis HC, Fragopoulou E and Pantazidou A (2002) Isolation and structural elucidation of biologically active phospholipids from *Scytonema julianum* (cyanobacteria). Biochem. J, 367 : 287-293.

Apte SK and Prabhavathi N (1994) Rearrangements of nitrogen fixation (*nif*) genes in the heterocystous cyanobacteria. J BioSci, 19(5) : 579-602.

Caire G, Storni de Cano MM, Palma RM and Zaccaro MC (2000) Changes in soil enzymes activity by cyanobacterial biomass and exopolysaccaride. Soil Biol.and Biochem.

Carrapico F, Teixeira G and Diniz MA (2000) *Azolla* as a biofertiliser in Africa-A challenge for the future. *Revista de Ciências Agrárias*, 23 (3-4): 120-138.

Chiu WL, Peters GA, Levieille G, Still PC, Couusins S, Osborne B and Elhai J (2005) Nitrogen deprivation stimulates symbiotic gland development in *Gunnera manicata*. Plant Physiol, 139:224–230.

Dawar S and Singh PK (2003) Disappearance of the symbiont *Anabaena azollae* in *Azolla* subjected to high phosphorus. IRRN, 28.1 : 53-54.

Drum RW and Pankratz S (1965) Fine structure of an unusual cytoplasmic inclusion in the diatom genus *Rhopalodia*. Protoplasma, 60 : 141-149.

Fergusson KM and Saint CP (2000) Molecular phylogeny of *Anabaena circinalis* and its identification in environmental samples by PCR. Appl Environ Microbiol, 66(9) : 4145-4148.

Fritsch C (1945) *The structure and reproduction of algae, vol II.* Cambridge Univ. Press, Cambridge.

Gentili F, Nilsson MC, Zackrisson O, DeLuca TH and Sellstedt A (2005) Physiological and molecular diversity of feather moss associative N_2-fixing cyanobacteria. J Exp Botany, cited online @ doi:10.1093/jxb/eri 309.

Guevara R, Armesto JJ and Caru M (2002) Genetic diversity of *Nostoc* microsymbionts fron *Gunnera tinctoria* revealed by PCR-STRR fingerprinting. Microb Ecology, 44 : 127-136.

Haroun SA and Hussein MH (2003) The promotive effect of algal biofertilizers on growth, protein pattern and some metabolic activities of *Lupinus termis* plants grown in siliceous soil. Asian J Plant Sciences, 2 (13) : 944-951.

Herdman M, Janvier M, Rippka R and Stanier RY (1979) Genome size of cyanobacteria. J Gen Microbiol, 111 : 73-85.

Kaneko T, Sato S, Kotani H, Tanaka A, Asamizu E, Nakamura Y, Miyajima N, Hirosawa M, Sugiura M, Sasamoto S, Kimura T, Hosouchi T, Matsuno A, Muraki A, Nakazaki N, Naruo K, Okumura S, Shimpo S, Takeuchi C, Wada T, Watanabe A, Yamada M, Yasuda M and Tabata S (1996) Sequence analysis of the genome of the unicellular cyanobacterium *Synechocystis* sp. strain PCC 6803. II. Sequence determination of the entire genome and assignment of potential protein-coding regions, DNA Res., 3 : 109-136.

Komárek J (2003) Coccoid and colonial cyanobacteria. In: Wehr JD and Sheath RG (eds), Freshwater Algae of North America. Ecology and Classification. Academic Press, Elsevier Science, USA. pp. 59-68.

Lin C, Liu ZZ, Zheng DY, Tang LF and Watanabe I (1989) Reestablishment of symbiosis to *Anabaena*-free *Azolla*. Sci. China Ser B 32 : 551-559.

Lindblad P and Costa JL (2002) The cyanobacterial-cycad symbiosis. Biology and Environment: Proceedings Royal Irish Academy, 102B (1) : 31–33

Meeks JC (1998) Symbiosis between nitrogen-fixing cyanobacteria and plants. BioScience, 48 : 266–276

Meeks JC and Elhai J (2002) Regulation of cellular differentiation in filamentous cyanobacteria in free-living and plant-associated symbiotic growth states. Microbiol Mol Biol Rev, 66(1) :94-121.

Meeks JC, Campbell EL, Summers ML and Wong FC (2002) Cellular differentiation in the cyanobacteria *Nostoc punctiforme.* Arch Microbiol, 178 : 395-403.

Mishra U and Pabbi S (2004) Cyanobacteria-A potential biofertilizer for rice. Resonance, June: 6-10.

Murakami A, Miyashita H, Iseki M, Adachi K and Mimuro M (2004) Chlorophyll d in an epiphytic cyanobacterium of red algae. Science, 303:1633.

Nakamura Y, Kaneko T, Sato S, Ikeuchi M, Katoh H, Sasamoto S, Watanabe A, Iriguchi M, Kawashima K, Kimura T, Kishida Y, Kiyokawa C, Kohara M, Matsumoto M, Matsuno A, Nakazaki N, Shimpo S, Sugimoto M, Takeuchi C, Yamada M and Tabata S (2002) Complete genome structure of the thermophilic cyanobacterium *Thermosynechococcus elongatus* BP-1. DNA Research, 9:123–130

O'Brien HE, Miadlikowska J and Lutzoni F (2005) Assessing host specialization in symbiotic cyanobacteria associated with four closely related species of the lichen fungus *Peltigera*. Eur J Phycol, 40:363–378

Oberholster PJ, Botha AM, Muller K and Cloete TE (2005) Assessment of the genetic diversity of geographically unrelated *Microcystis aeruginosa* strains using amplified fragment length polymorphisms (AFLPs). Afr J Biotechnol, 4(5):389-399.

Oren A (2004) A proposal for further integration of the cyanobacteria under the Bacteriological Code. International J Systematic Evolutionary Microbiol, 54:1895–1902

Osborne BA, Cullen A, Jones PW and Campbell GJ (1992) Use of nitrogen by the *Nostoc–Gunnera tinctoria* (Molina) Mirbel symbiosis. *New Phytologist* 120: 481–487.

Owe MC, Gantar M and Elhai J (1999) Reconstitution of a cycad-cyanobacterial association. Symbiosis, 27:125-134

Pabby A, Prasanna R and Singh PK (2004) Morphological characterisation of cultured and freshly separated cyanobionts (Nostocales, Cyanophyta) from *Azolla* species. Acta Botanica Hungarica, 46 (1-2): 211-223.

Rai AN and Bergman B (2002) Creation of new nitrogen-fixing cyanobacterial associations. Biol Environ:Proc Royal Irish Acad, 102B(1):65-68.

Rai AN, Soderback E and Bergman B (2000) Tansley Review: Cyanobacterium-plant symbioses. New Phytologist, 147:449-481.

Raven JA and Allen JF (2003) Genomics and chloroplast evolution:What did cyanobacteria do for plants. Genome Biol, 4(3):209.1-209.5.

Rippka R, Deruelles J, Waterbury JB, Herdmann M and Stanier RY (1979) Generic assignment, strain histories and properties of pure culture of cyanobacteria. J Gen Microbiol, 111:1-61.

Schopf JW (1996) Cyanobacteria. Pioneers of the early Earth. In: *Contributions in Phycology*. A.K.S.K, Prasad, J.A. Nienow and V.N.R Rao (eds), Nova Hedwigia, Beiheft 112, J.Cramer, Berlin, pp.13-32.

Singh SC, Sinha RP and Hader DP (2002) Role of lipids and fatty acids in stress tolerance in cyanobacteria. Acta Protozool, 41:297-308.

Steindler L, Huchon D, Avni A and Ilan M (2005) 16S rRNA phylogeny of sponge-associated cyanobacteria. Appl Environ Microbiol, 71(7):4127-4131.

Tang LF, Watanabe I and Liu CC (1990) Limited multiplication of symbiotic cyanobacteria of *Azolla* spp. on artificial media. Appl. Environ. Microbiol., 56(11):3623-3626.

Thajuddin N and Subramanian G (2005) Cyanobacterial biodiversity and potential applications in biotechnology. Curr Sci, 89(1):47-57.

Tirkey J and Adhikary SP (2005) Cyanobacteria in biological soil crusts of India. Curr Sci, 89(3):515-521.

Villarreal JCA and Renzaglia KS (2006) Structure and development of *Nostoc* strands in *Leiosporoceros dussii* (Anthocerotophyta): a novel symbiosis in land plants. American J Bot, 93:693-705.

Wilson JT (1980) In Blue green algae and Rice. PA Roger and SA Kulasooriya. International Rice Research Institute, Los Banos, Philippines.

Worden A, Chisholm SW and Binder BJ (2000) *In situ* hybridization of *Prochlorococcus* and *Synechococcus* (marine cyanobacteria) spp. with rRNA-targeted peptide nucleic acid probes. Appl Environ Microbiol, 66(1):284-289.

Mind Bogglers

1. How symbiosis by cyanobacteria with higher plants differs from rhizobial symbiosis?
2. What are the physiological changes in heterocysts and normal vegetative cell in filamentous cyanobacteria?
3. What is the role of cyanobacteria in sustainable agriculture?

4. What is the cause and significance of red colored glands of *Gunnera* in obtaining cyanobacterial symbiosis?
5. What are the major plant growth promoting compounds secreted by cyanobacteria that help in the growth and development of the inoculated plant?
6. Briefly describe the various techniques for mass production of BGA. What are the constraints of the use of cyanobacterial biofertilizers?
7. Genome sequencing of *Anabaena* has opened new ways to unravel the symbiotic interaction and development of better nitrogen fixing strains. Comment.
8. Give a brief note on the molecular techniques for characterization and identification of known as well as novel cyanobacteria.

4

Free Living, Loose Associative and Endophytic Diazotrophs

- Introduction 88
- Asymbiotic Nitrogen Fixers 89
- Associative Nitrogen Fixers 98
- Endophytic Nitrogen Fixers 102
- Future Perspectives 107
- References 108
- Mind Bogglers 113

Introduction

Diazotrophs are phylogenetically diverse microbes spanning over various classes varying vastly on basis of physiological properties (Young, 1992), include various phototrophic microorganisms aerobic phototrophic *Cyanobacteria* (Vaishampayan *et al*, 2001), anaerobic purple-sulfur phototrophs like *Chromatium*, and green-sulfur phototrophs, e.g. *Chlorobium*, *Rhodopseudomonas;* chemolithotrophs *Alcaligenes*, *Thiobacillus*, *Methanosarcina*, *Methanobacterium*, *Desulfovibrio* or *Azospirillum lipoferum* and heterotrophic bacterial anaerobes (*Clostridium*), microaerophiles (*Herbaspirillum*), and aerobes (*Azotobacter*). Free-living aerobic nitrogen fixing bacterial genera found in soil include obligate aerobes as *Azotobacter*, *Beijerinckia*, *Derxia*, *Achromobacter*, *Mycobacterium*, *Arthrobacter* and *Bacillus* but the majority are microaerophilic (e.g. *Azospirillum*, *Herbaspirillum*) or facultative and obligate anaerobes (e.g. *Klebsiella*, *Aerobacter*, *Desulfovibrios*, *Rhodospirillum*, *Clostridium*, *Erwinia*).

Asymbiotic class includes all diazotrophs that lack specific symbioses due to independent survival of these diazotrophs in terrestrial or aquatic environments and thus the part of nitrogen fixation by these counterparts is taken as asymbiotic nitrogen fixation (ANF). Being diverse, these diazotrophs could be categorized on basis of their life strategies. *Free-living* diazotrophs comprise the most commonly defined group of diazotrophic life strategies, grow in soil or on the soil surface or in associations with plant roots, excluding endophytic and symbiotic microorganisms and are termed as *asymbiotic* or *associative* diazotrophs on basis of degree of association. *Associative* diazotrophs show various degrees of plant association, from being truly endophytic to merely enriched presence in the rhizosphere or on root surfaces i.e rhizoplane.

BNF by free-living diazotrophs for agricultural purposes

- Caron (1900) reported yield increases following inoculation with saprophytic diazotrophs
- potential of soil to fix atmospheric N_2 was first described by Berthelot in 1888
- Winogradsky (1893) reported the isolation of the first heterotrophic diazotroph, *Clostridum pasteurianum*
- Gerlach and Vogel (1902) could not reproduce the positive effects
- Mishustin (1970) claimed that only a third of experiments reported statistically significant effects on yield.
- Okon (1995) reported positive results as 5-30% increase in crop yield in 60-70% of field-inoculations with *Azospirillum*

ASYMBIOTIC NITROGEN FIXERS

In majority ANF is primarily performed by heterotrophic free-living nitrogen fixing bacteria belonging to family Azotobacteriaceae that include three genera *viz. Azotobacter, Beijerinckia* and *Drexia* while another class of bacterial forms the free living cyanobacteria. A recent review by Kennedy and Islam (2001) reviews the contribution and potential of asymbiotic nitrogen fixation.

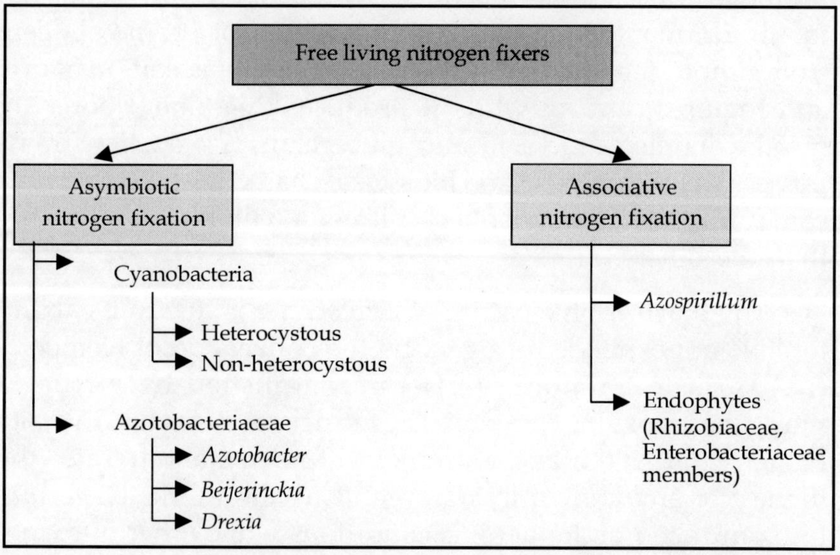

Types of free living nitrogen fixers on basis of extent of association

Azotobacter: As compared to other bacteria, *Azotobacter* is very large sized (ranges from 2-10x1-2.5 mm) gram negative, pleomorphic (different sizes and shapes) bacteria, produces copious amounts of polysaccharides and motile due to peritrichous flagella. Some species of this genus produce pigments, yellow-green, red-violet, or brownish-black while extracellular slime production is a common feature of genus, that helps in sequestration of water in the immediate vicinity and retain it.

Old cultures of *Azotobacter* exhibit presence of encapsulated forms 'cysts' that possess enhanced resistant to heat, desiccation and adverse conditions while germinate on advent of favorable conditions to form vegetative cells. *Azotobacter* sp. are sensitive to acidic pH, high salts, and temperature above 35^{0}C. This genus includes several species which could be distinguished on basis of cell size, pigmentation and extracellular slime production.

Cyst formation in *Azotobacter* is special character but this characteristic debars the bacterial nitrogen fixing potentials. Moreover these cysts do not behave as spore (for being not totally dormant). *Azotobacter* uses carbon for its metabolism from simple or compound substances of carbonaceous in nature and requires calcium while it produces thiamine, riboflavin, nicotin, indole acetic acid and gibberellin. *Azotobacter* strains possess varied chemical and biological characteristics equipping some strains to fix atmospheric nitrogen with higher efficiency than the others.

The diazotrophic nature of *Azotobacter* is due to its ability to "fix" atmospheric nitrogen, by the conversion of elemental (N_2) form to ammonia (NH_3) accomplished by virtue of nitrogenase enzyme complex. The nitrogenase enzyme diversity (Chien *et al*, 2000) and extremely rapid metabolic rate (the highest of any known living organism) allows this bacterium to fix nitrogen under aerobic conditions while other nitrogen-fixing bacteria possess only a single species of nitrogenase, which

General features of different species of the genera *Azotobacter*

Species	Origin/isolation	Flagellation	Extracellular slime production	Pigmentation
A.chrococcum	Neutral or alkaline Dutch soils	Peritrichous	Moderate	Black-brown insoluble pigment
A.agilis a.k.a. *Azomonas agilis*	Dutch (Delft) canal water	Peritrichous	Little to moderate	Green, fluorescent soluble pigment
A.vinelandii	New Jersy (USA) soils	Peritrichous	Little to moderate slime	Yellow, green, or violet fluorescent soluble pigment
A.insignis a.k.a *Azomonas insignis*	Small rivulets of the Cibodas Nature Reserve, Western Java, Indonesia	lophotrichous	Little	yellow-green, red-purple, or violet water soluble diffusible pigment
A.macrocytogen es a.k.a. *Azomonas macrocytogenes*	Danish soils	Polar	Abundant	Pink soluble pigment
A.paspali	Rhizosphere of *Paspalum notatum* variety "Batatais" (common Bahia grass) in Brazil,	Peritrichous	Little to moderate	yellow-green, fluorescent or red-violet, water soluble pigment
A.beijerinckii	North American soils	Absent	Moderate	yellowish or cinnamon pigment Yellow-light brown insoluble pigment
A.nigricans	East European soils	Absent	Moderate	brown-black or red-violet diffusible water soluble pigments
A.armeniacus	South American soils	Motile	Low	diffusible brown-black or red-violet water soluble pigment
Azotobacter salinestris	Slightly saline soils of western Canada and saline Egyptian soils	Motile peritrichous	variable	Absolutely dependent on Na(+) super ions for growth, N_2 fixation under microaerophilic, aeroadaptive conditions

*Three *Azomonas* species previously grouped as *Azotobacter* species differ from later, on basis of microcyst formation and % G+C content

requires anaerobic to microaerophilic conditions in order to function. This enhanced versatility of *Azotobacter* (regarding nitrogenase enzyme complex) makes the microbe attractive for agricultural purposes. The species of *Azotobacter* are known to fix on an average 10 mg of N/g of sugar in pure culture on a nitrogen free medium with maximum of 30 mgN fixed per gram of sugar. However, being poor competitor for nutrients in soil its utility as nitrogen biofertilizer becomes limited. Most efficient strains of *Azotobacter* would need to oxidize about 1000 kg of organic matter for fixing 30 kg of N/ha. In Indian soils, the population of *Azotobacter* is not more than 1x10^4 to 1x10^6 cfu/g^{-1} of soil. *Azotobacter* population is mostly influenced by other microorganisms present in soil, which may either stimulate *Azotobacter* population in soil thereby increasing the nitrogen fixation or adversely affect its population thus hampering nitrogen fixation by these bacteria. Punjab soils are characteristically low in organic compounds or active carbon status and moreover a variety of other microbes compete with *Azotobacter* for the active carbon.

Utility as nitrogen biofertilizer

Seed application of *Azotobacter* improves seed germination to a considerable extent and it also controls several plant diseases caused by bacteria, fungi and viruses due to production of a variety of compounds by the vegetative cells. The most common azotobacterial species that occur mostly in fertile soil is *A. chroococcum* however different strains vary in their nitrogen fixing ability which depends upon pH of soil, crop and other soil edaphic factors. The non-specificity of the bacteria and ability of free living nitrogen fixation is prime regarding its elaborate use in several kinds of crops and tree species. Several reports have depicted the benefit of *Azotobacter* inoculations in various crop plants. *Azotobacter* inoculation was reported to increase the yield attributes and nitrogen uptake in wheat (Singh *et al,* 1999). Narula *et al* (2005) have reported significant improvement in the cotton yield due to *Azotobacter* inoculation.

Organic supplementation with organic sources as manure, compost or vermiculite along with *Azotobacter* inoculation could further enhance crop yield as organic carbon rich soils show quite encouraging results than seed inoculation of *Azotobacter* of poor organic carbon soils that exhibit marginal benefits. Dual inoculations as in case with mycorrhizal or P-solubilizing microorganisms further benefit the crop plants in terms of enhanced root and shoot growth. Behl *et al* (2003) have reported the positive impact on root characters of wheat by *Azotobacter chrococcum* and mycorrhizal inoculation.

Now a days the trend of inoculation of mutant strains having enhanced nitrogenase or other engineered properties results in increased benefits to the inoculated crop. Lakshminarayana *et al* (2000) have obtained analogue-resistance mutants of *Azotobacter chroococcum* strain A103, derepression of nitrogenase enzyme and showed early ammonium excretion. Inoculation with these mutant strain in wheat caused increase in grain yield from 1.2-33.3%. Such mutants could be used as potential biofertilizer/inoculants for several cereal crops along with rice seedlings.

Beijerinckia: This free living, aerobic, chemoheterotrophic nitrogen fixer exhibits characteristic features i.e. formation of copious amount of slime, rod-shaped cells with round ends having polar lipoid bodies, nitrogen fixation at very acidic pH values (3.0–4.0) which differentiate this genera from other bacteria (*Azotobacter*).

Derxia: This genera contains a single species *D. gummosa* (Jensen *et al*, 1960) characterized by presence of pleomorphic, motile (by short polar flagellum) gram negative bacilli occurring singly or in short chains, capable of nitrogen fixation under aerobic or microaerobic conditions. Its efficiency of nitrogen fixation varies between 9 and 25 mgN/g of glucose consumed, but in most strains it is distinctly lower than in *Azotobacter* or *Beijerinckia*.

Free living cyanobacteria

Cyanobacteria have been reported from marine, subthermal vents, salt water or acid lakes to freshwater lakes, streams and rivers. These are microscopic prokaryotes comprising unicellular, colonial and filamentous types (form conspicuous large colonies or mats) (Rai, 1990) and are known to occur both in oxic and anoxic environments. Order Nostocales and Stigonematales members are the most common nitrogen fixers and include *Anabaena, Anabaenopsis, Aulosira, Chlorogloea, Cylindrospermum, Nostoc, Calothrix, Scytonema, Tolypothrix, Fischerella, Haplosiphon, Mastigocladus, Stigonema* and *Westiellopsis* as representative genera.

Bacterial species	**Location of isolation**	**Signature features**
Beijerinckia derxii	originally isolated from Australian soil	Yellow fluorescent pigment producing colonies, pigment production enhanced on iron deficient media, giant large cells
Beijerinckia derxii subsp. *Derxii*	Asian and South American soils	motile cells, utilizes nitrate
Beijerinckia derxii subsp. *Venezuelae*	Alkaline South American soils	Nonmotile, unable to utilize nitrate as sole source of N,
Beijerinckia fluminensis	originally isolated from "Baixada Fluminensis" Rio de Janeiro, Brazil	Older cultures show characteristic large capsules enclosing 2–10 or more individual cells
Beijerinckia indica	isolated from soils of all continents in tropical and few subtropical regions	-Lipid bodies persist in aged cultures, No resting bodies (cyst or ascococcus) formation
Beijerinckia indica subsp. *indica* (Type species)	-do-	-cells lack flagella
Beijerinckia indica subsp. *Lacticogenes*	-do-	butyrous and brittle colonies, production of small amounts of lactic acid
Beijerinckia mobilis	Isolated from very acid tropical soils and in soil samples from Southeast Asia (Indonesian (Javanese) soils), Africa, and South America	Vegetative cells motile, Ascococcus-like clusters of cells are often visible in older cultures

There are two general types *viz* the coccoid and filamentous types. The coccoid species occur as single cells, colonies or masses of various shapes with cell arrangement in rows to form flat plate, radial arrangement in spherical colonies. The colonies in mat remain enclosed in a gelatinous sheath that varies in consistency and thickness. The filamentous types generally possess the nitrogen fixing attribute which is characteristic of heterocystous cyanobacteria but exceptions to this have been reported where non-heterocystous types exhibited nitrogen fixation. Filamentous cyanobacterial cells are characteristically arranged in rows to form trichomes that are then enveloped in a common enclosing sheath to form filaments.

Some filamentous species exhibit cell differentiation and form *heterocysts*, characterized as large cells heterotrophic in nature, possessing extraordinarily thick cell wall connected by cytoplasmic bridges to neighboring photosynthetic cells, lack an oxygenic photosystem and lack biliprotein pigments and carboxysomes thus appearing empty amidst normal pigmented vegetative cells. Heterocysts provide congenial environment for effective functioning of nitrogenase and are considered as the sites of nitrogen fixation. The number of heterocysts could be taken as a rough parameter to indicate the nitrogen fixing capacity of cyanobacteria as amount of nitrogen fixed depends upon physiological and environmental factors. Apart from forming heterocysts some heterocystous cyanobacterial species form akinete cell type that germinates on advent of favorable conditions to form new vegetative cells for multiplication and formation of a new colony.

As stated above, the cyanobacterial ecology advocates their occurrence in almost all ecotypes or geological conditions. This versatility may be accounted to their unique feature to form akinete or other dormant structures for survival and then their high adaptability index to withstand harsh abiotic conditions. Being originally classified among algae due to their microscopic morphology, pigmentation and oxygen evolving photosynthesis (photosystems PS II and PS I connected in series) these now are classified as prokaryotes. The genome size of cyanobacteria, representative of all major taxonomic groups, lies in the range, $1.6X10^9$ to $8.6X10^9$ Da which is comparable to that of other bacteria (1.0 to $3.6X10^9$ Da) (Herdman *et al,* 1979).

Biochemistry and genetics of nitrogen fixation in heterocystous cyanobacteria

The nitrogenase activity in isolated heterocysts of *Anabaena variabilis* depends upon supply of H_2 while both on O_2 and H_2 under dark conditions. The nitrogen fixation results in formation of glutamine amino acid which is the chief product of this activity. The formed glutamine is then translocated to vegetative cells and this translocation depends on transfer of glutamate from the latter cell types. Algal cells contain high levels of glutamine sythetase (GS) and low levels of glutamine oxoglutarate amino transferase (GOGAT) which further supports the fact that heterocysts import glutamate from vegetative cells and then add NH_4^+ to it to form glutamine. Heterocyst differentiation could be repressed or controlled by glutamine or a metabolite of glutamine other than ammonium ions.

Heterocysts lack PSII (photosystem II), which is involved in O_2 evolution and ferredoxin reduction however it contains the PSI system that can also reduce ferredoxin. The reduced ferredoxin can donate electrons to nitrogenase and thereby provides the electrons required by this process. Strikingly heterocysts also contain uptake hydrogenase that fine tunes the loss of electrons by nitrogenase system thus increasing its efficiency.

Nitrogenase of BGA is similar to that other nitrogen fixing bacteria as it catalyzes reduction of molecular nitrogen to NH_4^+ ions, requires ATP, Mg^{2+} and reductant, is oxygen sensitive and can also catalyze the reduction of protons, cyanide and C_2H_2. Moreover the Mo-Fe and Fe-protein of *Anabaena cylindrica* and *Plectonema boryanum* can cross react with the Fe-proteins of *Azotobacter* and *Clostridium.*

Utility as biofertilizer

The inherent fertility of tropical rice field soils could be attributed to the activity of N_2-fixing cyanobacteria. A variety of cyanobacterial strains colonize rice fields wherein heterocystous species are capable of fixing atmospheric nitrogen. However, several non-heterocystous cyanobacteria are able to fix atmospheric nitrogen under microaerophilic conditions.

In situ estimations using acetylene reduction technique have shown an addition of 18–15 kgN ha^{-1} yr^{-1} due to the activity of diazotrophic cyanobacteria. The role of N_2-fixing cyanobacteria in maintenance of the fertility of rice fields has been well substantiated and documented all over the world. In India alone, the beneficial effects of cyanobacteria on yield of many rice varieties have been demonstrated in a number of field locations (Dhaliwal et al, 1995). Beneficial effects of cyanobacterial inoculation have also been reported on a number of other crops such as barley, oats, tomato, radish, cotton, sugarcane, maize, chilli and lettuce. The use of algae and cyanobacteria in waste treatment is beneficial in different ways since they can bring about oxygenation and mineralization, in addition to serving as food source for aquatic species. Using the marine cyanobacteria *Oscillatoria* sp. BDU 10742, *Aphanocapsa* sp. BDU 16 and a halophilic bacterium *Halobacterium US* 101, Uma and Subramanian (1990) treated ossein factory effluent which resulted in reduced calcium and chloride levels and enabled 100% survival and multiplication of *Tilapia* fish with only cyanobacteria as feed source.

Overcoming oxygen sensitivity of algal nitrogenase

Heterocystous cyanobacteria

- Structural modification of some vegetative cells to thick walled heterocysts
- Heterocysts lack PS II that is actively engaged in O_2 evolution by photolysis of water
- Presence of uptake hydrogenase
- High activity of oxidative pentose pathway

Non-heterocystous cyanobacteria

- Compartmentalization/separation of photosynthesis and nitrogen fixation
- Presence of antioxidants
- O_2 linked hydrogen uptake
- Presence of internal membrane systems that serve in intracellular O_2 protection

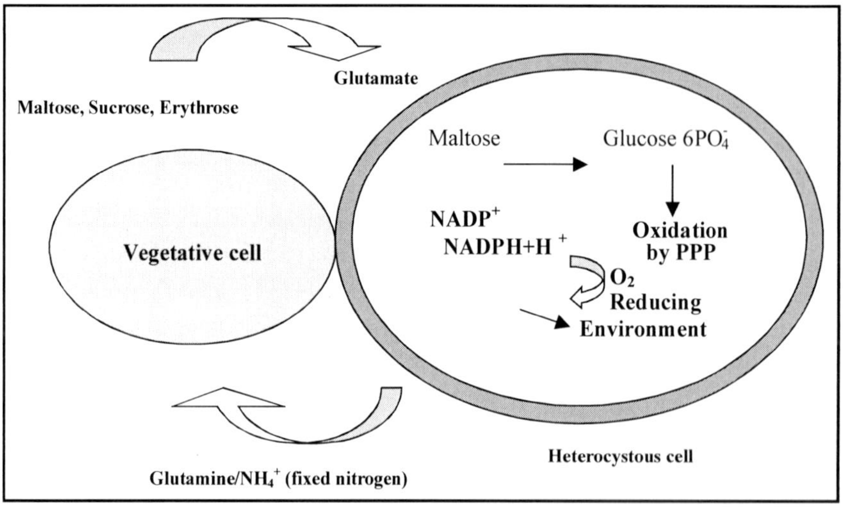

Formation of glutamate and its transpotation to the neighboring vegetative cells

Associative Nitrogen Fixers

Bacteria are major inhabitants of rhizosphere, rhizoplane and histoplane and play a vital role in pronounced plant growth by increasing the availability of nutrients to the plants as well as impart positive effect in terms of regeneration of the quality of soil (Alami *et al*, 2000).

Azospirillum: Azospirillum belongs to class α-proteobacteria comprised of plant associated diazotrophs. It resides in close association with plant roots and exhibit loose associative/ endophytic nitrogen fixation in C_4 plants. It is a gram-negative, aerobic bacteria that fixes nitrogen under low concentration of dissolved oxygen (<8 mM) using a high-affinity terminal oxidase and behaves as a typical microaerophilic N_2-fixer.

Seed/seedling inoculation with this diazotroph results in increased colonization of lamellae present on base of root hair or in the area of longitudinal contact with/between epidermal cells and this enhanced colonization results in increasing the number of root hairs, their length and density (Patriquin *et al*, 1983). Moreover both root length and surface area increases, probably due to release of phytohormones by the bacteria (auxin IAA) synthesized by several trytophan-dependent IAA biosynthetic pathways (Zakharova *et al*, 1999) and Indole-3-acetamide (IAM) pathway through a Trp-independent route (Prinsen *et al*, 1993).

Pellicle formation and change in colour from light green to blue due to growth of *Azospirillum* in 0.17% Doberenier's medium

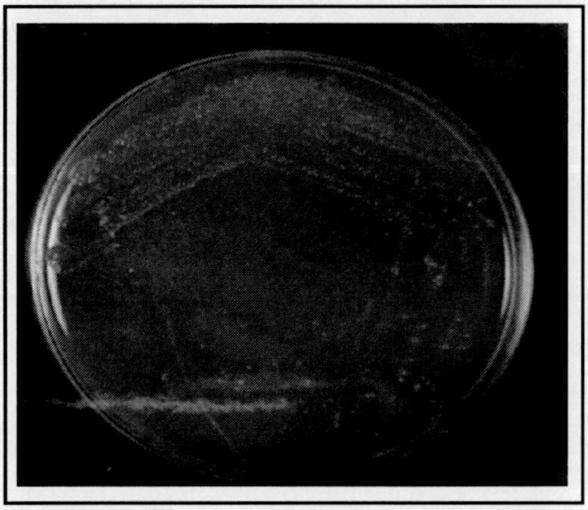

Azospirillum on Doberenier's agar

(For a coloured version of this figure, see plate section, page 300)

Very low pectinolytic activity was detected in extracts of wheat roots inoculated with *Azospirillum* (Kapulnik *et al*, 1984) which may be involved in slight hydrolysis of middle lamellae bacteria colonized cortex cells without causing cell collapse and may accelerate uptake of water and minerals by roots. Thus *Azospirillum* inoculation could result in increased mineral (N,

P, K and other microelements) uptake and improved water status of plants and can result in increased accumulation of dry matter and thus increased yields in fields. Didonet *et al* (2000) have reported the reallocation of N and biomass in wheat inoculated with *Azospirillum sp.*

The establishment of the azosprilla to host plant roots is important for the creation of conducive and efficient association and this is performed by *Azospirillum* cells by virtue of certain specific gene products. *A. brasilense* cells are able to aggregate under certain environmental conditions, leading to the formation of bacterial flocs, a property that affects the dispersal and survival of these bacteria in soil. The attachment requires involvement of extracellular proteins and polysaccharides basically the outer membrane proteins (OMPs) in aggregation and root adsorption of *A. brasilense* as these proteins have been observed to function as adhesions in gram negative bacteria as in case of *Pseudomonas fluorescens* and *Rahnella aquatilis* equipping these rhizobacteria to cling onto plant roots (Achouak *et al,* 1995). A similar OMP gene has recently been isolated, sequenced and identified in *A. brasilense* that codes for 367 amino acid purified protein (Mr~38,753 Da) that exhibits significant homology to certain bacterial porins as well as the sequence exhibits the presence of cell surface exposed domain involved in adhesion process (Burdman *et al,* 2001).

Effect on various crop plants: Inoculation of azospirilla in plants results in positive effects (significant morphological and physiological changes) on plant growth and yield of many crops of agronomic importance probably by improved water uptake and mineral utilization. Rueda-Puente *et al* (2004) have reported an increase in plant length as well as increased in total protein, lipid and ash content of the two pickleweed (*Salicornia bigelovii*) genotypes inoculated with *Azospirillum halopraeferens* and *Klebsiella pneumoniae.* Bhaskara Rao and Charyulu (2005) have reported the increase in various plant growth parameters *viz.* plant height, dry weight of shoot and root, and total N content of shoot, root and grain yield of foxtail millet (variety Lepakshi) due to inoculation by three strains of *Azospirillum lipoferum* either alone or in combination with nitrogen fertilizer (40 kg N ha^{-1}). The grain yield of the inoculated plants was significantly higher compared to the control plants. Inoculation of *Arospirillum* also resulted in increase in panicle and seed weight. Similarly munis *et al* (2003) have reported growth stimulatory effects of inoculation of *Arospirillum* on wheat and *Vigna radiata*.

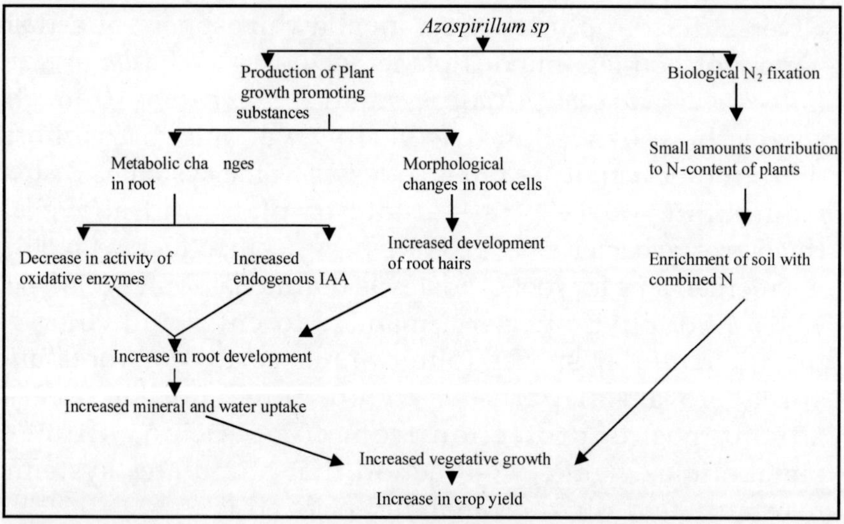

Mechanisms of *Azospirillum sp* resulting in increased growth and yield of crop plants

Increase in root biomass of *Azospirillum* inoculated maize plants

(For a coloured version of this figure, see plate section, page 300)

Endophytic Nitrogen Fixers

As discussed in the introduction of the chapter, diazotrophs could be categorized on basis of spatial and functional relationship between microbes and host plant into following broad groups *viz.* Exophytic; when diazotroph remain outside the host plant and Endophytic; when diazotroph occur within the internal tissues of the host plant. Diverse diazotrophic bacteria inhabit rhizosphere of a wide variety of non-leguminous plants including *Klebsiella oxytoca, Enterobacter cloacae, Alcaligenes* and *Azospirillum* (from the rhizosphere of wetland rice). However the rhizospheric diazotrophic activity is prone to losses as the system is distant from main source of assimilate of plant and above all rhizospheric bacteria have to face heavy competition with other microorganisms for root exudates as nutritional source (Quispel, 1991). Endophytic bacteria manage to enter and colonize the intercellular spaces (outer/inner cortex) of roots and spread to aerial parts *via* xylem thus cherish several advantages as protection from competition, intimate metabolic exchange, PGP action that stimulates systemic resistance and direct availability of fixed N.

Due to their inherent ability to colonize inside the plant roots endophytic diazotrophs have become the glaring arena in BNF technology. The renewed interest in endophytic diazotrophs, such as *Acetobacter, Azoarcus,* and *Herbaspirillum,* in gramineous plants has arisen because of their occurrence mainly within plant tissues and evidence for significant nitrogen fixation (James and Olivares, 1998).

Azoarcus sp. from Kallar grass abundantly colonizes and expresses *nif* genes and nitrogenase protein inside the original host as well as in rice roots (Reinhold-Hurek and Hurek, 1998). *Acetobacter diazotrophicus,* an endophytic diazotroph from sugarcane, has been found mainly associated with sugar-rich plants, such as sugarcane, sweet potato and Cameroon grass (Baldani *et al,* 1997). Diazotrophic *Herbaspirillum seropedicae* has been isolated from plant parts of maize, sorghum, sugarcane, and other gramineous plants . The endophytic populations of a variety of plants are affected by a number of biological and environmental factors such as plant cultivar, plant age, tissue type, time of sampling, and soil environment.

At the cytological level only a few investigations have been worked out to characterize endophytic bacterial infection and colonization (Taechowisan *et al*, 2003). Histological studies involving the immunoflorescence or confocal scanning electron microscopy have provided relevant information on the route followed for infection by the bacteria as well as the location of multiplication and survivality. Largely two categories of endophytic diazotrophs could be delineated on basis of the bacterial genera inhabiting the internal tissues *viz*, rhizobial and non rhizobial

Rhizobial endophytic diazotrophs

These include various rhizobial/species or strains residing in the cortical and xylem tissues of a variety of non-leguminous plants. Few examples have been given in different crops.

a) Rhizobium

Recent years have witnessed the extension of rhizobial inoculations to several non leguminous crops and the benefits rendered by these bacteria to these crops. Rhizobial inoculations particularly the exopolysaccharide producing strains were reported to have immense positive impact on root adhering soil (RAS) per root biomass as well as soil macropore volume, which resulted in increased withstanding capacity under water deficit conditions as well as effective utilization of applied nitrogen fertilizer in *Helianthus annuus* L (Alami *et al*, 2000). However the rhizobial cells function as traditional endophytic diazotroph in majority of non legumes and provide the atmospheric nitrogen which is assimilated more closely than the one fixed by rhizospheric diazotrophs.

b) Bradyrhizobium

It was reported to occur in roots of wheat and wetland rice-*Oryza breviligulata* which grows in same wetland site as *Aeschynomene,* an aquatic stem nodulated legume (Chaintreuil *et al,* 2000). This gram negative, aquatic, photosynthetic, slow growing bacteria fixes nitrogen under free living conditions and secretes phytohormones like IAA and gibberellic acid. It is reported to colonize the rice roots by root cap colonization, intercellular invasion of epidermal cells or via intracellular invasion of the root tissues. The mechanism followed by this bacteria to invoade rice roots is via crack entry invasion mechanism which include movement by polar attachment, loose root cap cells, fissures caused by protruding lateral roots.

c) *Azorhizobium*

It is gram negative, aerobic, small, motile rod, isolated from root and stem nodules of tropical legume *Sesbania rostrata* (Kannaiyan *et al,* 2001). This bacteria posseres few characteristics that help it to perform the function of an endophytic diazotroph; such as it fixes N under free living conditions, tolerates upto 12mM dissolved O_2, secretes extracellular lipopolysaccharides and exhibits symbiotic route in *S. rostrata* and associative endophytic diazotrophic route in rice. Further it invades the roots of inoculated plants via crack entry at point of lateral root emergence, intercellular invasion in outer/inner cortex and colonizes cortical tissue and xylem as these locations particularly xylem vessels provide a suitable nutritional and environmental niche owing to low pO_2 and site for exchange of metabolites necessary for nitrogen fixation (O'Callaghan *et al,* 1997).

The bacteria were also found associated with paranodules which are nodular structures produced by modification from lateral roots due to stimulating effect of auxins such as 2,4-D (Kennedy *et al,* 1997; Kalia and Gupta, 2002) under laboratory conditions.

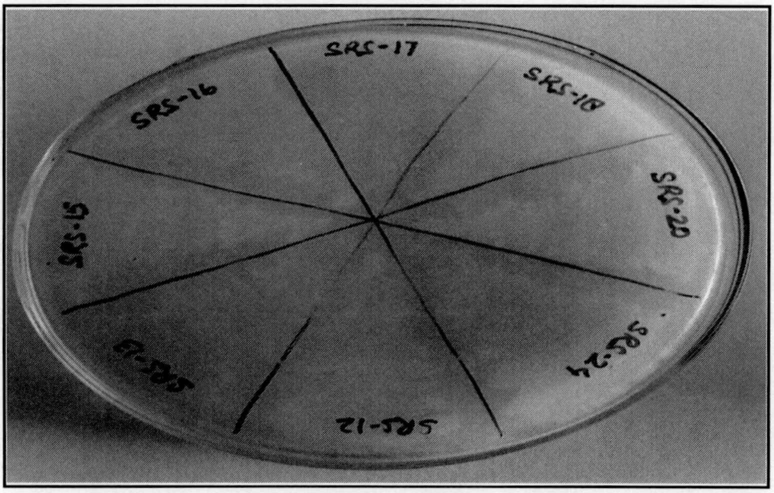

Acid production by azorhizobial isolates depicted by yellowing of BTB supplemented YEMA media

(For a coloured version of this figure, see plate section, page 301)

Non Rhizobial endophytic diazotrophs

d) Gluconacetobacter diazotrophicus

Gluconacetobacter diazotrophicus (syn. *Acetobacter diazotrophicus*)-sugarcane association represents a model system for monocot-diazotrophic associations. This allows experimentation to answer questions pertaining to their establishment, colonization process, biological nitrogen fixation, growth promotion etc. It was observed to contribute substantially large amount of biologically fixed nitrogen to sugarcane crop almost half of fixed N (Sevilla *et al,* 2001). However it could also be isolated from other plants *viz.* cameroon grass, sweet potato, coffee, tea, banana, ragi, rice and insects that infest sugarcane like pink mealybugs.

G. diazotrophicus is a gram negative, acid tolerant, obligate aerobe, straight rod with curved ends, occurring in pairs/chains, non-endospore former, opportunistic endophyte isolated from roots and stems of sugarcane and not root free rhizospheric soils. It is highly saccharophilic as grows on 10% sucrose concentration and at low pH (3.0), secretes extracellular levansucrase, is antagonistic to

Colletotrichum falcatum, and is able to solubilize phosphorus (Muthukumarsamy *et al*, 2002). Measurements of the acetylene reducing activity of propagating stem cuttings suggested that nitrogen-fixing bacteria (species then unknown) moved from sugarcane stem cuttings to roots and the new stems (Patriquin *et al*, 1983). This bacterium persists from one generation to the next of the vegetatively propagated plant (Dong *et al*, 1997).

It also invades via crack entry at point of lateral root emergence, loose root cap cells, tear wound sites and leaves i.e. damaged stomata. It colonizes the root and stem xylem vessels and may also occur intercellularly in apoplastic stem spaces of stem cortex and phloem sieve tubes (Boddey *et al*, 2003). Alam *et al* (2003) have observed a comparable germination, tiller number and cane yield at zero and 120 kg N ha^{-1} due to presence and contribution of nitrogen via BNF by gram negative bacterium isolated from root extracts and cane juice.

e) *Herbaspirillum*

Herbaspirillum sp. is an endophytic diazotrophic bacteria associated with important agricultural crops. It is isolated from roots of maize, rice, soyabean, forage grasses and recently from leaves, stems and roots of sugarcane (Olivaries *et al*, 1996). *Herbaspirillum* cells are gram negative, aerobic, curved, motile rods that secrete enzymes as pectinase and cellulase. This endophytic diazotroph invades roots by adsorption and proliferation on secondary roots and root surface damage. It colonizes intercellular spaces in young tertiary leaves and coleoptiles and prefers apoplast of shoot tissues (Elbeltagy *et al*, 2001). Baldani *et al* (2000) observed positive effect of inoculation of selected strains of endophytic diazotrophs *Herbaspirillum seropedicae* and *Burkholderia* spp under gnotobiotic conditions in rice plants that showed medium and high increases in their fresh weight in response to inoculation while the contribution of the strains to plant N via BNF varied from 31 to 54% and also increase N content in grains of inoculated plants.

f) Azoarcus

It is a gram negative, microaerobic, strictly respiratory bacillus bacteria that secretes cellulolytic enzymes and is isolated from roots or stems of rice, wheat and kallar grass. It invades roots via crack entry, adsorption at lateral root emergence and colonizes cortex in elongation zone above root tip both intra and interacellularly and penetrate upto stele of roots (Hurek *et al*, 1997).

Future Perspectives

Today's world is teeming with people concerned about the conservation and judicious/sustainable utilization of resources and thus the trend is increasing of going organic including higher marketing and consumption of organically grown products. The asymbiotic and associative/endophytic diazotrophs are gaining popularity regarding their dual benefit of providing nitrogen to the plant as well as increasing the growth parameters by actively releasing/providing plant growth promoting substances to the inoculated plant. The non-specificity of their action spectrum is the most beneficial facet equipping these diazotrophs to be used for a variety of plant species. However the problems of variable effects on yield and loss of fixed nitrogen due to denitrification have to be sorted out which profoundly debar their proficient use and prominence among farmers.

Use of immobilized bacterial strains on agrarose beads or polyurethane foam decrease the chances of flushing out of these cells from the rhizospehric region. Moreover making the plants more responsive as well as increase in nitrogen use efficiency of the inoculated plants would be a step towards better management of the nitrogen fixed diazotrophically. Genetic manipulations using the pectinase/cellulase genes in certain endophytes enhance their ability to colonize the internal plant root tissue.

New methods are being employed to study the establishment of the inoculated bacteria so that colonization at enhanced rate/significant population on or in the host roots, particularly under competitive conditions in nonsterile substrates could be deduced (Ma *et al,* 2001). Several modern techniques of *in situ* detection of endophytic isolates without rigorous culturing have been observed to be useful regarding economy of time. Certain other limitations of culture-based methods invite use of molecular methods of microbial community analysis (Gamalero *et al,* 2003). Molecular biological techniques, such as phylogenetic probes or PCR-based approaches, allow the detection of particular DNA sequences and therefore can be used to trace their presence in target organisms directly in the environment and may also involve RNA/protein based methods for detection of diversity of asymbiotic/endophytic diazotrophs in a given rhizospheric soil/plant sample. The genetic analysis includes deciphering the functional ORFs of *nif* gene clusters among different isolates. However the use of reporter genes for marking the qualitative presence of inoculated bacteria is most popular particularly for the establishment studies of endophytic diazotrophs (Lowder and Oliver, 2001; Ma *et al,* 2001). These techniques are more sophisticated and save a lot of time required by traditional culturing techniques.

References

Achouak W, De Mot R and Heulin T (1995) Purification and partial characterization of an outer membrane protein involved in the adhesion of *Rhanella aquatilis* to wheat roots. *FEMS Microbiol Ecol,* 16:19-24.

Alam R, Karim MZ, Fida Hasan M, Hossain MA and Mafizur Rahman ABM (2003) Evaluation of sugarcane clones with biological nitrogen fixation endowment. *Asian J Plant Sciences,* 2(2):216-219.

Alami Y, Achouak W, Marol C and Heulin T (2000) Rhizosphere soil aggregation and plant growth promotion of sunflowers by an exopolysaccharide-producing *Rhizobium* sp. strain isolated from sunflower roots. *Appl Environ Microbiol,* 66(8):3393-3398

Baldani JI, Caruso L, Baldani VLD, Goi S and Dobereiner J (1997) Recent advances in BNF with non-legume plants. Soil Biol. Biochem., 29:911-922

Baldani VLD, Baldani JI and Döbereiner J (2000) Inoculation of rice plants with the endophytic diazotrophs *Herbaspirillum seropedicae* and *Burkholderia* spp Biology Fertility Soils, 30(5-6): 485 - 491

Behl RK, Sharma H, Kumar V and Narula N (2003) Interactions amongst mycorrhiza, *Azotobacter chroococcum* and root characteristics of wheat varieties. *J Agronomy Crop Sci,* 189(3): 151-155.

Bhaskararao KV and Charyulu PBBN (2005) Evaluation of effect of inoculation of *Azospirillum* on the yield of *Setaria italica* (L.). *African J Biotechnol,* 4(9):989-995.

Boddey R M, Urquiaga S, Alves BJR and Reis V (2003) Endophytic nitrogen fixation in sugarcane: present knowledge and future applications. *Plant Soil,* 252: 139-49.

Burdman S, Dulguerova G, Okon Y and Jurkevitch E (2001) Purification of the major outer membrane protein of *Azospirillum brasilense,* its affinity to plant roots, and its involvement in cell aggregation. *MPMI,* 14(4):555–561.

Caron A (1900) Management of the Ellenbach estate. *Jahr Deut Landw Gesell,* 15:43.

Chaintreuil C, Giraud E, Prin Yves, Lorguin J, Ba A, Gillis M, deLajudie P and Dreyfus B (2000) Photosynthetic Bradyzhirobia are natural endophytes of the African wild rice *Oryza breviligulata. Appl Environ Microbiol,* 66:5437-47.

Chien YT, Auerbuch V, Brabban AD and Zinder SH (2000) Analysis of genes encoding an alternative nitrogenase in the archaeon *Methanosarcina barkeri* 227. *J Bacteriol,* 182: 3247-3253.

Didonet AD, Lima O, Candaten AA and Rodrigues EO (2000) Reallocation of nitrogen and biomass to the seeds in wheat inoculated with *Azospirillum* bacteria. *Pesq Agropec Bras,* 35(2):401-411.

Dong Z, McCully ME and Canny MJ (1997) Does *Acetobacter diazotrophicus* live and move in the xylem of sugarcane stems? Anatomical and physiological data. *Annals of Botany,* 80:147-158.

Elbeltagy A, Nishioka K, Sato T, Suzuki H, Ye B, Hamada T, Isawa T, Mitsui H and Minamisawa K (2001) Endophytic colonization and in planta nitrogen fixation by a *Herbaspirillum* sp. isolated from wild rice species. *Appl Environ Microbiol*, 67(11):5285-5293.

Gamalero E, Lingua G, Berta G and Lemanceau P (2003) Methods for studying root colonization by introduced beneficial bacteria. *Agronomie*, 23:407–418

Gerlach M and Vogel I (1902) Stickstoffsammelnde Bakterien. *Zeitschrift Bakterien II*, 8:669.

Dhaliwal MK, Pandher Ms. Gupta RP, Garcha HS and gagnija MS (1995). Effect of Chemical nitrogen on the growth and nitrogen fixation by blue green algar in Basmati rice. *Ind. J. Ecol.* 22(1): 7-10.

Herdman M, Janvier M, Rippka R and Stanier RY (1979) Genome size of cyanobacteria. *J Gen Microbiol*, 111:73–85.

Hurek T, Wagner B and Reinhold-Hurek (1997) Identification of N_2 fixing plant and fungus-associated *Azoarcus* species by PCR-based genomic fingerprints. *Appl Environ Microbiol*, 63: 4331-39.

James EK and Olivares FL (1998) Infection and colonization of sugarcane and other graminaceous plants by endophytic diazotrophs. *Crit Rev Plant Sci*, 17:77-119

Jensen HL, Petersen EJ, De PK and Bhattacharya R (1960) A new nitrogen-fixing bacterium: *Derxia gummosa* nov. gen. nov. spec. *Archiv fur Mikrobiologie*, 36:182-195

Kalia Anu and Gupta RP (2002) Nodule induction in non-legumes by rhizobia. *Ind J Microbiol*, 42:183-193.

Kannaiyan S, Kumar K and Gopalaswamy G (2001) Stem nodulating *Sesbania rostrata* as a bio-resource for rice production, induction of nodulation and endophytic N_2 fixation in rice. *Madras Agric J*, 88:547-60.

Kapulnik Y, Gafny R and Okon Y (1984) Effect of *Azospirillum spp.* Inoculation on root development and nitrate uptake in wheat (*T.aestivum* cv. Miriam) in hydroponic systems. *Can J Bot*, 63 : 627-631.

Kennedy IR and Islam N (2001) The current and potential contribution of asymbiotic nitrogen fixation to nitrogen requirements on farms: a review. *Aus J Exptl Agricul*, 41:447-457.

Kennedy IR, PeregGerk LL, Wood C, Deaker R, Gilchrist K and Katupitiya S (1997) Biological nitrogen-fixation in non-leguminous field crops: Facilitating the evolution of an effective association between *Azospirillum* and wheat. *Plant Soil*, 194:65-79.

Lakshminarayana K, Shukla B, Sindhu S S, Kumari P, Narula N and Sheoran R K (2000) Analogue-resistant mutant of *A.chrococcum* derepressed for nitrogenase activity and early ammonia excretion having potential for inoculants for cereal crops. *Ind J Exp Biol*, 38(4): 373-378.

Lowder M and Oliver DJ (2001) The use of modified GFP as a reporter for metabolic activity in *Pseudomonas putida*. *Microb Ecol*, 41:310–313.

Ma W, Zalec K and Glick BR (2001) Biological activity and colonization pattern of the bioluminescence-labeled plant growth-promoting bacterium *Kluyvera ascorbata* SUD165/26. *FEMS Microbiol Ecol*, 35:137–144.

Mishustin EN (1970) The importance of non-symbiotic nitrogen-fixing microorganisms in agriculture. *Plant Soil*, 32:545.

Munir A, Munir I, Afrasyab S and Hasnain S (2003) Growth stimulatory effects of *Azospirillum* strains on *Triticum aestivum* and *Vigna radiate. Biotechnol*, 2(3):198-205.

Muthukumarasamy R, Revathi G, Seshadri S and Lakshminarasimran (2002) *Gluconacetobacter diazotrophicus* (syn. *Acetobacter diazotrophicus*) a promising diazotrophic endophyte in tropics. *Current Science*, 83:137-45.

Narula N, Saharan BS, Kumar V, Bhatia R, Bishnoi LK, Lather BPS and Lakshminarayana K (2005) Impact of the use of biofertilizers on cotton (*Gossypium hirusitum*) crop under irrigated agro-ecosystem. *Arch Agron Soil Sci*, 51:69-77.

O'Callaghan KJ, Davey MR and Cocking EC (1997) Xylem colonization of the legume *Sesbania rostrata* by *Azorhizobium caulinodans. Proc R Soc Lond B*, 264:1821-26.

Okon Y, Itzigsohn R, Burdman S and Hampel M (1995) Advances in agronomy and ecology of the *Azospirillum*/plant association, *In*

Nitrogen Fixation:Fundamentals and Applications, IA Tikhonovich, NA Provorov, VI Romanov and WE Newton (ed.), Kluwer Academic Publishers, Dordrecht, p. 635-640.

Olivares FL, Baldani VLD, Reis VM, Baldani JI and Dobereiner J (1996) Occurrence of the endophytic diazotrophs *Herbaspirillum* spp. in root, stems, and leaves, predominantly of Gramineae. *Biol Fertil Soils,* 21:197-200

Patriquin DG, Dobereiner J and Jain DK (1983) Sites and processes of association between diazotrophs and grasses. *Can J Microbiol,* 29:900-915.

Prinsen E, Costacurta A, Michiels K, Vanderleyden J and Van Onckelen H (1993) *Azospirillum brasilense* indole-3-acetic acid biosynthesis: evidence for a non-tryptophan dependent pathway. *Mol Plant Microb Interact,* 6:609-615.

Quispel A (1991) A critical evaluation of the prospects for Nitrogen fixation with non legumes. *Plant Soil,* 137:1-11.

Rai AN (1990) Cyanobacteria in symbiosis. Handbook of Symbiotic Cyanobacteria (ed. Rai, AN), CRC Press, Boca Raton, 1990.

Reinhold-Hurek B and Hurek T (1998) Life in grasses: diazotrophic endophytes. *Trends Microbiol,* 6:139-144

Rueda-Puente EO, Castellanos T, Toryo-Dieguez E and Diaz de Leon-Alvarez JL (2004) Effect of *Klebsiella pnemoniae* and *Azospirillum halopraeferens* on the growth and development of two *Salicornia bigelovii* genotypes. *Aus J Exptl Agricul,* 44:65-74.

Sevilla M, Burris RH, Gunapala N and Kennedy C (2001) Comparison of benefit to sugarcane plant growth and $^{15}N_2$ incorporation following inoculation of sterile plants with *Acetobacter diazotrophicus* wild-type and *nif* mutant strains. *Mol Plant Microbe Interact,* 14:359-366.

Singh G, Somani L L and Totawat K L (1999) Scope of supplementing nitrogen requirement of wheat through organic source and *Azotobacter. Internatl J Tropical Agricul,* 17(1-4): 177-182.

Taechowisan T, Peberdy JF and Lumyongs (2003) Isolation of endophytic actinomycetes from selected plants and their antifungal activity. *World J. Microbiol Biotechnol,* 19: 381-385.

Uma L and Subramanian G (1990) Effective use of cyanobacteria in effluent treatment. *In Proceedings of the National Symposium on Cyanobacterial N_2 Fixation*, IARI, New Delhi, 1990, pp. 437–444.

Vaishampayan A, Sinha RP, Hader DP, Dey T, Gupta AK, Bhan U and Rao AL (2001) Cyanobacterial biofertilizers in rice agriculture. Botanical Revs, 67:453-516.

Winogradsky S (1893) Sur l'assimilation de l'azote gazeux de l'atmosphere par les microbes. Comptes Rendus Academie Sciences, 116:1385-1388.

Young JPW (1992) Phylogenetic classification of nitrogen-fixing organisms, p. 43-86. *In* G. Stacey, R. H. Burris, and H. J. Evans (ed.), Biological Nitrogen Fixation. Chapman & Hall, New York, NY.

Zakharova EA, Shcherbakov AA, Brudnik VV, Skripko NG, Bulkhin NS and Ignatov VV (1999) Biosynthesis of indole-3-acetic acid in *Azospirillum brasilense*. Insights from quantum chemistry. Eur. J. Biochem. 259, 572-576.

Mind Bogglers

1. How loose associative and endophytic nitrogen fixers are superior to free living nitrogen fixers?
2. What are the benefits and drawbacks of inoculation with *Azotobacter*? Which type of plant species or genera are most benefited by inoculation of the azotobacterial inoculum?
3. What is the reason of association of *Azospirillum* with C_4 plants? Is it possible to isolate these bacteria from cereal root damaged at the root tip region and why?
4. Endophytic diazotrophs share a common infection/ invasion pathway. How it differs from the classical rhizobial invasion system in legumes? Illustrate with a labeled diagram.
5. The nitrogen fixing potential of free living cyanobacteria is generally and largely dependent on the heterocystous filamentous types. Comment. Are there any other types

of cyanobacteria involved in BNF? If yes enlist the genera and also provide the mechanism(s) by which they debar oxygenic inactivation of nitrogenase system.

6. The oxygen paradox in nitrogen fixers is most elaborately complex yet versatile problem. Comment on the statement. How *Azotobacter* is most versatile and equipped in combating oxygen inactivation of nitrogenase complex?

7. How far is it authentic to comment that endophytic diazotrophs could be instrumental in extending legume-rhizobia like symbiotic association to cereal/non-leguminous crops? What are the major problems in extending symbiotic association in non-legumes?

8. Why and how different strains of the same genera/species differ in their nitrogen fixing capabilities?

5

Phosphate Solubilizing Microorganisms

- ❐ Introduction 116
- ❐ Phosphorus Reserves in Soil 117
- ❐ Phosphate Solubilizing Microorganisms 119
- ❐ Methods for Detection of P-solubilizing Ability 122
- ❐ Mechanism of P-solubilization 122
- ❐ PSM P-solubilization Mechanism 124
- ❐ Organic Acids Produced by PSMS 124
- ❐ Genetics of Phosphate Mineralization 125
- ❐ Potential Uses in Agriculture 128
- ❐ Future Perspectives 130
- ❐ References 131
- ❐ Mind Bogglers 136
- ❐ Appendix 137

INTRODUCTION

Increased population pressure, reduced length of fallow, deforestation and improper agricultural practices have led to widespread soil degradation in many parts of the developing world which has culminated in appearance of determinative manifestations of environmental damage in form of inadequate replenishment of soil nutrients and organic matter. Phosphorus is second major plant growth limiting nutrient and despite being in abundance in soil (organic and inorganic form), P deficiency is becoming critical in many soils throughout world as these soils have free available P concentration (generally) not higher than 10mM even at most soluble slight acidic pH (Arnou, 1953). Moreover, consistent P deficiency threatens to disturb the use efficiency of applying other nutrients. Hence to preserve the sustainability of agriculture it is foremost to rebuild soil fertility and maintain/improve the soil biodiversity to reap hiked current levels of productivity and farm income.

P deficiency is alleviated by the massive application of P_i fertilizers, currently estimated to be about 40 million metric tons per year worldwide (Plaxton, 2004). Only 10-25 % of phosphorus, introduced by the way of fertilizers is acquired by plants while the major portion of introduced P, as a result of chemical changes in soil, is transformed to insoluble and is hard to reach for plants forms as well as a substantial part is flushed out by rains to land aquifers. It demands new additions of fertilizers and, thus, there is an accumulation in soil of insoluble compounds of phosphorus ("over phosphatization" of soils). The situation is aggravated by vast application of complex fertilizers, keeping apart from nitrogen and potassium resulting in accumulation and excess of one on expense of lack of other elements. Such imbalance influence both quality and harvest as a whole.

P reserves are also finite and based on current rate of use, it is expected that the world's known reserves of high quality rock phosphate will be depleted within the current century (Isherwood, 2000). This situation clearly demands the application of specific measures or microbes for sustainable presence and availability of labile P pool in soil.

Phosphorus reserves in soil

Phosphorus exists in soil in six different forms (Holford, 1989), which in order of increasing stability are:

1) P in the soil solution (< 1% of the total)
2) inorganic P (Pi) in the plant residues
3) inorganic P adsorbed on surfaces of colloidal mineral particles (clay) and organic matter
4) inorganic P in various compounds, which range from sparingly soluble to extremely insoluble
5) inorganic P occluded or adsorbed in P reactive minerals
6) very stable organic P in plant, animal and microbial residues

Insoluble soil phosphates are oxyapatite, hydroxyapatite, carbonate apatite and fluoroapatite in neutral and alkaline soils and hydroxyphosphates of iron and aluminum as dufreenite, strengite and avriate in acidic soils (Subba Rao, 1995). The organic soil P pool particularly the soil microbial biomass may account from 2 to 5% of the total P and around 5 to 10% of soil organic P in agricultural soils, or 10-100 kg P/ha in top 10 cm of soil, which is equivalent to or greater than the amount contained in plant material growing in the soil. In general, forest, grassland and some tropical soils have large amount of organic form of P. Thien and Myers (1991) indicated that by increasing soil microbial activities, bioavailability of P in a bioactive soil was remarkably enhanced. The facts that certain soil microbes are capable of dissolving relatively insoluble phosphatic forms (Bojinova *et al*, 1997) has opened the possibility for inducing microbial solubilization of phosphates in soil.

However a variety of factors affect the organic phosphorus content of a soil such as:

- soil with higher clay content contains greater amount of organic P
- high rainfall generally decreases organic phosphorus content and poor drainage of soils develops conducive environment for higher organic P levels
- decline in soil pH generally increases the organic P content
- decrease in organic P content due to cultivation
- phosphorus content of the parent material
- soil sulphur content required for increase in microbial biomass
- type of microorganisms present which will affect the rate of mineralization and immobilization

More recent studies have found that the rate of P-flux through the microbial biomass is, in fact, considerably greater (Oberson *et al*, 2001). P in microorganisms that occurs predominantly in organic forms (or as polyphosphates) is rapidly mineralized and is readily available for uptake by plant roots (Macklon *et al*, 1997) but this availability differs according to soil physico-chemical characters as well as the soil foodweb components i.e. micro and macroorganisms.

Plant available phosphorus: Phosphorus is absorbed by plants primarily as the monovalent phosphate ion ($H_2PO_4^-$) and less rapidly as the divalent phosphate ion ($H_2PO_4^{2-}$), with the soil pH controlling the relative abundance of the two forms and among these 50% of the total dissolved P present in soil solution may be organic forms of P (Richardson, 1994). The general availability of soluble orthophosphate ions exist as $H_2PO_4^-$ $>H_2PO_4^{2-}>PO_4^{3-}$. At low pH of 4 to 5 there occurs relative abundance of $H_2PO_4^-$ ions while decrease in acidity of soil increases $H_2PO_4^{2-}$ ions followed by PO_4^{3-}. Phosphate availability to plants depends on soil pH (Dave and Patel, 2003). At high pH, P get fixed on reaction with calcium while at lower pH it reacts with iron and aluminum ions (Jakasaniya and Trivedi, 2004). As the soluble P is utilized, it is replenished from the labile P pool, at a rate regulated by:

- quantity of P in the labile pool, and the soils buffering capacity (sorbtivity). Buffering capacity is the ability of soil to maintain its P concentration in solution as living organisms absorb P, or as P is added by fertilization (Holford, 1989). The higher the soils buffering capacity, the slower the P comes into solution.
- root morphology, including total root length, abundance of root hairs, rate of root growth, and the total volume of soil occupied by roots (Richardson, 1994)
- kinetics of P uptake by roots. P mineralizing phosphatase enzyme activity is decreased by high phosphate concentrations. This activity has been observed to be considerably higher in the rhizosphere than bulk soil and strongly correlated with total microbial activity.

PHOSPHATE SOLUBILIZING MICROORGANISMS

Occurrence: Microorganisms, being one of the component of soil foodweb, play several important roles as decomposition, nutrient mobilization/mineralizaton processes, storage of water and soil nutrient pool in form of microbial biomass, biological nitrogen fixation as well as denitrification reactions (Whitelaw, 2000). Among the above microbial diversity certain microbes are involved in solubilization of plant unavailable inorganic/organic phosphorus pool and thus convert insoluble phosphatic compounds to soluble forms (Raju and Reddy, 1999) and are termed as P solubilizing microorganisms (PSMs). PSM diversity spans over both bacterial as well as fungal microbial types classified to different families and representative genera involved are (appendix) *Pseudomonas, Bacillus, Micrococcus, Flavobacterium, Rhizobium, Bradyrhizobium, Enterobacter* and *Synechocystis* (PSBs) while *Penicillum, Fusarium, Curvularia, Trichoderma, Fomitopsis* and *Aspergillus* are the major P solubilizing fungal genera (Kang *et al,* 2002). Apart from these genera certain cyanobacteria and actinomycetes are also able to solubilize the unavailable inorganic P resources of soil. The increase in plant growth characteristics and crop yield is partly due to enhanced P availability as most of active PSMs are known to produce plant growth promoting substances of various types ranging from phenolics to protein products. However, the presence of phosphobacteria in sufficient quantities is always necessary in soil for obtaining higher inoculation efficacy on crop plants (Subba Rao, 1982).

The phosphate solubilizing microorganisms are observed to be exclusively in higher number in the rhizosphere of the host plant than the bulk soil thus making rhizosphere to be the most active site for P transformations (Kundu *et al*, 2002). Phosphate solubilizing (PS) organisms account for 26-39% of the microbial population of the rhizosphere of wheat, ryegrass and sub clover (Kucey *et al*, 1989). There is also some evidence to show that legumes support greater numbers of PS organisms than do non-legumes, and similarly some legume plants support greater numbers of PS microorganisms than other legumes.

Phosphorus is an essential component of very essential macromolecules *viz.* phophorylated energy storing compounds (ATP, NADP), DNA and RNA. However it is also required for other functions in plants as:

- cell division, fat and albumin formation
- flowering and fruiting, including seed formation
- crop maturation
- root development, particularly of the lateral and fibrous rootlets
- strength of straw in cereal crops
- improvement of crop quality, especially of forages and vegetables

Screening, isolation and characterization : PSMs particularly PSBs could be isolated from rhizospheric soils by serial dilution plating on solid Pikovskaya's media (Pikovskaya, 1948) containing tri or bicalcium phosphates (TCP) and thus screening is performed to isolate microbes (bacterial or fungal colonies) able to solubilize TCP to form characteristic halos around growth. Apart from the above media slight modifications as addition of a pH sensitive dye like bromophenol blue could easily depict the observable changes in pH of the media towards acidic range that provides yellow colouration to the halo around the colony growth (Mehta and Nautiyal, 2001). A new media National Botanical Research Institute's phosphate growth medium (NBRIP) broth has been formulated by Johri *et al* (1999).

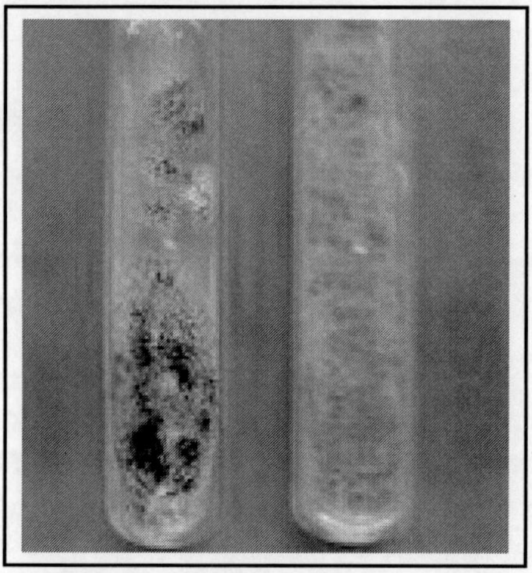

Solubilization of tricalcium phosphate in Pikovskaya's media by *Aspergillus niger* and *Trichoderma viridae*

The colonies that sustain the P solubilization attribute after six to seven subculturings could be further characterized both phenotypically as well as genotypically. The phenotypic characterizations include the use of basic microbiological techniques as morphological (includes shape, size and arrangement of cells, their gram reaction, endospore formation, motility), cultural (growth on solid/liquid media containing P source) and biochemical protocols (include an array of tests as carbohydrate utilization, H_2S production, MR/VP etc). Now a days several commercial kits are available that help in easier, faster and accurate identification and characterization of P solubilizers such as BIOLOG system that can identify a wide range of soil bacteria (BIOLOG, California, USA; http://www.biolog.com), API 20NE system (bioMérieux, France; http://www.biomerieux.com) allows the identification of species among the pseudomonads group (including *Acinetobacter*) and the IdentBacillus system (Mikrokit Ibérica S.L., Spain; http://www.bme.es/microkit) is able to differentiate between bacilli species.

Methods for detection of P-solubilizing ability

Physical tests that predict the amount of soluble P under broth culture conditions could also give the inkling about the P-solubilizing capability of a particular isolate. Modifications of the common physical method could be employed for indirect determination of solubilized phosphate in medium through spectrophotometric determination of residual TCP (Mikanova and Novakova, 2002). Casanova *et al* (2002) have reported the use of nuclear and related techniques for agronomic evaluation of phosphate fertilizers and the share of P supplied by PSMs to the plants. This work was supported by the research report given by Orlova (2003) regarding crystallographic properties of phosphates.

However the paradox of unculturable bacteria demands for the use of novel genotype oriented molecular techniques to characterize and pinpoint the presence of very complex and scarcely known microbial soil diversity. The molecular techniques involve the PCR based protocols among which the phylogenetic profiling uses PCR amplified rRNA oligotyping while others include TWO primer-Random amplified polymorphic DNA (TP-RAPD) protocol, staircase PAGE or staircase electrophoresis (SCE), rep-, ERIC- and BOX-PCR (include use of specific conserved short intergenic repeated sequences) (Igual *et al,* 2001).

Mechanism of P-solubilization

Up to 30% of cultivated soil microorganisms are able to mobilize sparingly soluble soil P sources (Mueller *et al,* 1997). Moreover, a modified composition of the sugar fraction in root exudates in response to P deficiency may increase the microbial production of P-mobilizing compounds in the rhizosphere (Schilling *et al,* 1998). Also the mineralization of various forms of organic soil P strongly depends on soil-microbial activity but significant amounts of soil-solution Pi can be sequestered into microbial biomass.

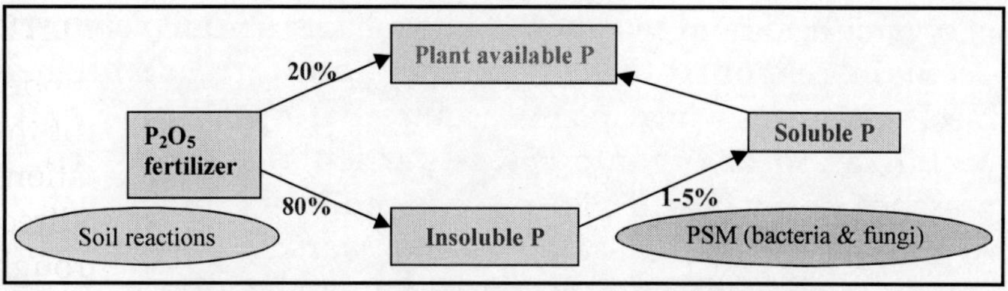

Increased P fertilizer use efficiency by PSMs

Microorganisms directly affect the ability of plants to acquire P from soil through a number of structural or process-mediated mechanisms. These include

(i) an increase in surface area of roots by either an extension of existing root systems (eg, mycorrhizal associations) or by enhancement of root branching and root hair development (i.e, growth stimulation through phytohormones),

(ii) by displacement of sorption equilibria that results in increased net transfer of phosphate ions into soil solution or an increase in the mobility of organic forms of P

(iii) through stimulation of metabolic processes that are effective in directly solubilizing and mineralizing P from poorly available forms of inorganic and organic P. These processes include the excretion of hydrogen ions, the release of organic acids, production of siderophores and production of phosphatase enzymes that are able to hydrolyse soil organic P reserves

P-solubilizing microorganisms are capable of excreting organic acids such as citric acid and oxalic acid, thereby enhancing dissolution of sparingly-soluble P sources, including Fe and Al phosphates and apatite, by means of acidification and complexation processes. Some microorganisms in variable-charge soils also demonstrate a greater capability of utilizing and transforming adsorbed P than the others (He *et al,* 2002). Hence processes such as rhizosphere acidification, exudation of organic acids and secretion of phosphatases from plant roots occur in response to P deficiency, and are established mechanisms by which plants acquire P (Randall *et al*, 2001). Mobilization of organic P was particularly expressed in presence of other acid phosphatase-producing fungi such as *Aspergillus fumigatus* (Tarafdar and Marschner, 1995). The mobilisation of sparingly soluble P sources by release of large amounts of organic chelators, such as oxalate seems to be mainly restricted to ECM fungi (Clark and Zeto, 2000) and may be not only involved in P acquisition but also in weathering of mineral soils and rocks (Van Breemen *et al*, 2000).

PSM P-SOLUBILIZATION MECHANISMS

- *Organic acid production:* PSM growth is generally accompanied by decrease in pH of medium due to production of organic acids that dissolute insoluble phosphates.
- *Siderophores:* These iron chelating compounds produced by PSBs and certain PGPR bacteria chelate iron from iron phosphates and thereby increases P availability in acid soils
- *Humic substances:* Humic and fulvic acids with carboxylic, phenolic, hydroxylic or phenol hydroxyl moieties release P from various P sources as compost, rock phosphate due to formation of stable complexes with calcium, iron and aluminінum.
- *Mineral acids:* Oxidation of reduced sulphur compounds by sulphur oxidizing bacteria (*Thiobacillus ferrooxidans, Thiobacillus thiooxidans*) results in formation of sulphuric acid or oxidation of pyrite to release P from insoluble phosphates
- *Proton extrusion:* Concomitant release of protons during respiration by H^+ ATPase (present at cell membrane) extrudes H^+ or during NH_4^+ assimilation H^+ are exchanged by uptake of NH_4^+.
- *Carbondioxide:* rhizospheric microbes during respiration release CO_2 which is converted to carbonic acid decreasing pH of microenvironment that helps in P solubilization.

ORGANIC ACIDS PRODUCED BY PSMS

- citric, glycolic, fumaric, malic, oxalic, succinic, tartaric, malonic, lactic, 2-ketogluconic, gluconic, glyoxalic, α-ketobutyric acid
- aliphatic acids being more effective than aromatic acids

- ❐ dibasic and tribasic acids being more effective than monobasic and aromatic acids
- ❐ organic acids act as chelating agents as
  - oxalic, citric, 2-ketogluconic and tartaric acids chelate calcium in calcareous/alkaline soil
  - formation of stable organometallic complexes with iron, alumininum or calcium as the central or conjugate cation decreases precipitation of solubilized phosphates

GENETICS OF PHOSPHATE MINERALIZATION

Some gram-negative bacteria have the ability to dissolve poorly soluble calcium phosphates such as hydroxyappatite and rock phosphate ore, a phenotypic behavior termed as mineral phosphate-solubilizing (Mps$^+$) phenotype. Action of Mps$^+$ bacteria on poorly soluble calcium phosphates in soils can enhance the availability of Pi for microbial and/or plant growth. *Erwinia herbicola* and *Pseudomonas cepacia* exhibit strong Mps$^+$ phenotype due to gluconic acid-mediated dissolution of the calcium phosphate. This mineral phosphate-solubilizing phenotype could be utilized as the basis for a renewable phosphate fertilizer direct oxidation pathway as well as one can speculate about the possible ecophysiological role of these genes in gram negative bacteria.

Phosphate mineralization is attributed to presence of two specific genes in Mps$^+$ phenotype *viz.* the gluconic acid dehydrogenase (GDH) that converts glucose to gluconate and the cofactor pyrroloquinoline quinone (PQQ) synthase genes (forms a peculiar quinoprotein required for Mps$^+$ phenoype). The PQQ biosynthesis genes have been isolated and cloned from P solubilizing *Erwinia herbicola* and the essential cofactor was observed to be product of two genes i.e. PQQ synthase (*pqqD*) and a putative PQQ transporter (*pqqE*) (Goldstein, 1995). The current model is that apo-GDH is integrated into its location on the outer face of the cytoplasmic membrane. When cofactor PQQ is present in the medium, it binds to apo-GDH to form the holozyme, which results in formation and secretion of gluconic acid and thereby solubilization of hydroxyappatite and rock phosphate. *E. coli* cells do not possess *mps* genes but *E.coli* K-12 strain having the constructed plasmid containing a functional mineral phosphate-solubilizing gene from *P. cepacia* E-37 has been shown to constitutively synthesize apo-GDH but not PQQ (Babu-Kahn *et al,* 1995).

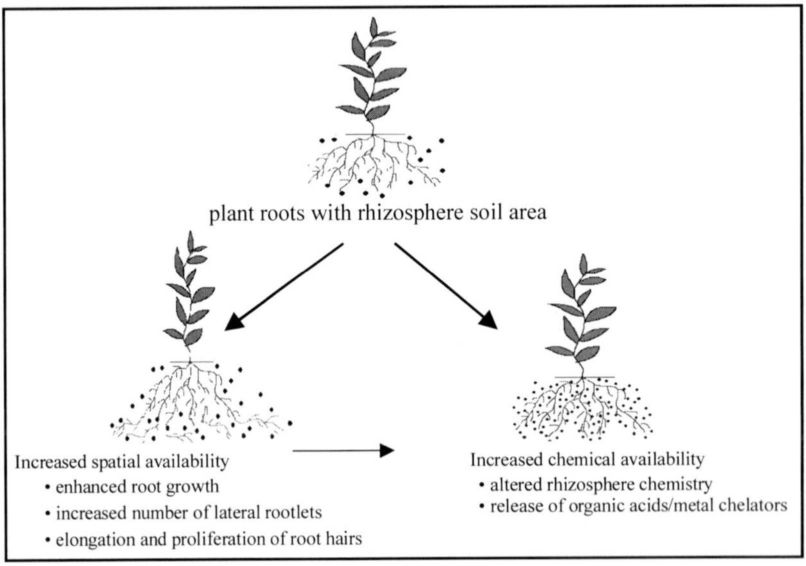

Mechanism of mineral acquisition from rhizosphere soil by plant roots in presence of PSMs or mycorrhiza

If functional PQQ synthesis genes from another bacterium are supplied, GDH activity is seen. The expression of this gene on a plasmid in *E. coli* JM109 or HB101 resulted in both dissolution of exogenous hydroxyapatite and production of gluconic acid. The completely sequenced gene (designated *gabY*) containing only one apparent ORF which has been reported to have a molecular mass of 14,235 kDa GabY. Disruption of *gabY via* site-directed mutagenesis results in loss of both the Mps^+ phenotype and the ability to produce gluconic acid. GabY has no apparent homology with putative PQQ synthase genes but does share regions highly homologous with the histidine permease system membrane- bound protein HisQ as well as other proteins in this family.

Genetic engineering protocols for development of effective PSMs particularly PSBs relies on the genetic information regarding production of organic acids as these *mps* genes could be gathered in rhizosphere competent bacterial strains that are more competent in colonizing the plant rhizosphere. The upper edge in inoculation of these strains is regarding presence of carbon source in form of root exudates as well as better utilization of solubilized P before its precipitation as cation phosphate salts. The action spectrum of already known strains could also be increased by over-expression of PQQ biosynthesis and GDH genes. Formation of dual inoculant by incorporation of PQQ genes in *Rhizobium* that carries the apo-GDH copy of the *mps* genes would be several times better owing to increased nitrogen fixation by P solubilizing actions (Gynaeshwar *et al*, 2002). Further obtaining transformants with unique ability to utilize novel and abundant carbon sources in the rhizosphere would help overrule the problem of competition of inoculant strain with native rhizosphere microbial population and thus increased establishment. An abundant and relatively common carbon source in rhizosphere is ACC (ethylene precursor) and engineering the competent cells to utilize this carbon source could further enhance the ability of inoculated PSBs to get colonized and established in the rhizosphere. Similarly other common carbon sources abundant in rhizosphere include the salicylate, organic herbicides or pesticides, which could help better establishment of the bateria over the natural native microflora relying on traditional carbon sources.

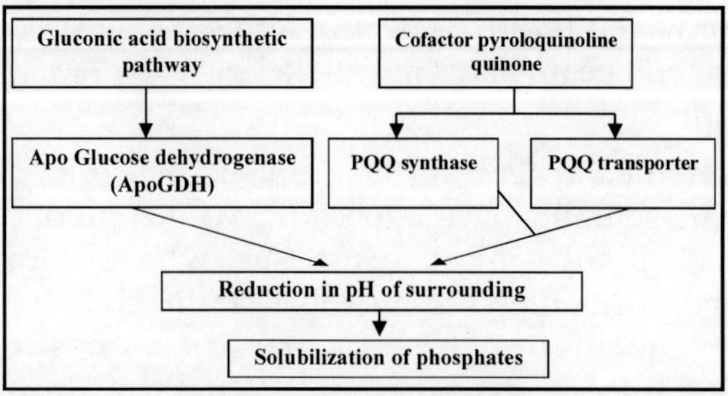

Molecular genetics of phosphate solubilizing microbial action

POTENTIAL USES IN AGRICULTURE

The PSMs have higher maximal transport rates of P than the ubiquitous soil microorganisms and the use of these microbes enhances the plant available P pool in soil thus enabling plant roots to absorb this essential element from rhizospheric soil region (Radhakrishnan, 1996; Prabhakaran et al, 1996). The reports of positive effect of inoculation of PSMs on legume as well as non legume crops have been enriched with the new research work in this field, however, the popular use of these microbes to reap visible profits is yet to reach its log phase (Gupta et al 1998). Several formulations are being launched by the private companies but little information is available regarding these products being par excellence from the farmer's view point.

Mehana and Wahid (2002) studied the effect of interactions between three phosphate dissolving fungi (PDF) (*Aspergillus niger, A. fumigatus* and *Penicillium pinophilum*) and *Rhizobium leguminosarum* biovar *viciae* (RH) on some soil chemical properties, phosphorus and nitrogen uptake and yield of faba beans (*Vicia faba* L.) and reported significant increase in soil available P level and seed and straw yield after inoculation with PDF and RH than untreated control as well as 50% reduction in recommended SP field rate in the presence of PDF. Gaur *et al* (1980) have also advocated the positive effect of inoculation of P-solubilizing bacteria on several crops due to enhanced availability of P nutrition. A similar report by Kucey (1987) provided positive effect in terms of increased P uptake by wheat and bean plants inoculated by a PSF *Penicillium bilaji* and AM fungi.

Combined direct application of rock phosphate and phosphate-solubilizing microbes (PSM) has produced mixed results on plant growth responses, which were perhaps attributed to differences in microbial strains and/or soils being treated. Inoculation of PSM for Rock Phosphate (RP) solubilization or reacting the RP with a liquid culture supernatant (LCS) of the PS fungi may be considered a better means to overcome the low solubility problems of Rock phosphate (Goenadi, 1996). Such an approach may eliminate factors inhibiting a successful interaction between PSM and RP under field conditions. Goenadi *et al* (2000) have observed that by using 9 day old LCS of phosphate-solubilizing fungus, *Aspergillus niger* BCC F.194 isolated from tropical acid soils on Morrocan rock phosphate (MRP), the P solubility was recorded to be highest (2.5 gL^{-1} MRP) thus decreasing the dependency on concentrated H_2SO_4 treatment.

Treatment	Effect of inoculation	Reference
Lab studies		
9 day old liquid culture supernatant of phosphate-solubilizing fungus (PSF), *Aspergillus niger* BCC F.194	P solubility of Morrocan rock phosphate was recorded highest (2.5 g L^{-1} MRP) thereby decreasing dependency on conc. H_2SO_4 treatment.	Goenadi *et al* (2000)
Pseudomonas fluorescens, *Bacillus megaterium* strains	*in vitro* solubilization of Ca_3PO_4 on solid and in liquid PVK media with pH reduction of 2.8 and 1.2 respectively	Alveraz (1991)
Moraxella sp	goethite, kalonite, montmorillonite P-complex utilization on agar plates observed	He *et al* (2002)
Pot studies		
Phophobacteria and mycorrhiza	Yield and yield attributes of wheat affected significantly	Raja *et al* (2002)
Azospirillum, B.megaterium, P.fluorescens	Increase by 37 and 53% of shoot dry weight, total nitrogen (N) yield and the shoot phosphorus content compared to the plants inoculated with *A. lipoferum* and uninoculated ones in wheat respectively	El-Komy (2005)
Field studies		
Bacillus megaterium var. *phosphaticum* (Phosphobacterin)	Yield due to inoculations for one third field trials for various crops	Mishustin and Naumova (1962)
Bacillus polymyxa, Pseudomonas straita	Inoculation along with RP increased soybean significantly over control	Sundara Rao (1968)
Phosphobacteria and mycorrhiza	Grain yield of wheat at par with full P level of PSB inoculated plots	Raja *et al* (2002)

Aspergillus niger, *A. fumigatus* and *Penicillium pinophilum*, *Rhizobium leguminosarum* biovar *viciae*	A. Significant increase in soil available P level and seed and straw yield after inoculation of faba beans (*Vicia faba* L.)	Mehana and Wahid (2002)
Rhizobium, Azospirillum Phosphobacteria in groundnut.	B. 50% reduction in recommended SP field rate	Balamurugan (1994)
Combined direct application of rock phosphate and phosphate-solubilizing microbes (PSM)	C. Mixed results on plant growth responses D. Eliminate factors inhibiting a successful interaction between	Goenadi (1996)

PSM inoculants are not popular to an extent demanded by the role these microbes play in supplying the reserve inorganic soil pools of phosphates. The basic constraint being their variability in providing the desirable impact at a quantitatively visible scale which largely depends on an array of soil edaphic factors including abiotic forces (soil composition, texture, pH, temperature, moisture content, substrate availability or organic matter content) and biotic factors (root exudates amount and quality, physiological status of inoculated strain(s), competition with native microflora, predation by microbial grazers as microarthropods and nematodes). In general the population sizes of introduced PSM strains declines rapidly upon introduction in soils, which could be overted by use of rhizosphere competent strains of PSMs.

FUTURE PERSPECTIVES

PSMs have been demonstrated to replenish the available form of P in soil from the rich insoluble pool of rich P reserves but limited success has been gained in attempts to capitalize on microbial processes to increase plant access to P from soil and/or fertilizer sources. Their use could only debacle on grounds of consistent commercial profits regarding either hiked grain yield or low P fertilizer input or increased P fertilizer use efficiency. A deeper understanding of the P mobilization processes and the interaction of the P solubilizers with other biotic and abiotic factors of soil rhizosphere i.e. ecological interaction and profile of the inoculant are required to be deciphered before acception or discarding of this group of microbial element mineralizers. There is dire need for isolation and estimation of better performing region specific local strains as well as regarding development of transconjugants with dual action particularly a nitrogen fixer along with enhanced P solubilization capacity. The presence and working of inoculant requires assurance in form of certain novel techniques or set of techniques that indicate them *in situ/in vivo.* Moreover use of PSFs liquid supernatant culture suspension could help in solubilizing rock phosphate thus decreasing sulphuric acid treatment dependency. Recent developments in microbial community analysis that do not rely on cultural procedures will provide better understanding of how microorganisms interact in complex environments as the phylogenetic profiling or rRNA oligotyping using staircase electrophoresis-PAGE and PCR based protocols are one of most elegantly sophisticated procedures for detection and visualization of specific microorganisms in the rhizosphere (Ranjard *et al*, 2000, Theron and Cloete, 2000).

References

Alvarez IK (1991) Quantification of solubilizadores phosphate microorganisms ground of deposit rock phosphoric Fresco the Mount. *Rev. Agron Face (Maracay)*, 17:373-379.

Arnou DI (1953) In Soil and Fertilizer Phosphorus in Crop Nutrition (IV) ed Pierre WH and Norman AG, Acad. Press, NY.

Babu-Khan S, Yeo TC, Martin WL, Durn MR, Rogers RD and Goldstein AH (1995) Cloning of a mineral phosphate-solubilizing gene from *Pseudomonas cepacia. Appl Environ Microbiol,* 61(3):972-978.

Balamurugan S (1994) Studies on the synergistic effect of *Rhizobium* spp with *Azospirillum* and phosphobacteria in groundnut (*Arachis hypogea* L.). M.Sc. (Ag.) Thesis, Tamil Nadu Agricultural University, Coimbatore.

Bojinova D, Velkova R, Grancharov I and Zhelev S (1997) The bioconversion of Tunisian phosphorite using *Aspergillus niger*. *Nutr Cyc Agroecosyst,* 47:227–232.

Casanova E, Salas AM and Toro M (2002) The use of nuclear and related techniques for evaluating the agronomic effectiveness of phosphate fertilizers, in particular rock phosphate. Monitoring mycorrhizas and phosphate solubilizing microorganisms, Tec-Doc 1272, IAEA, Vienna.

Clark RB and Zeto SK (2000) Mineral acquisition by arbuscular mycorrhizal plants. *J Plant Nutr*, 23:867-902.

Dave A and Patel HH (2003) Impact of different carbon and nitrogen sources on phosphate solubilization by *Pseudomonas fluorescens*. *J Microbiol*, 43:33-36.

El-Komy HMA (2005) Coimmobilization of *A. lipoferum* and *B. megaterium* for plant nutrition. *Food Technol Biotechnol*, 43(1):19-27.

Goenadi DH (1996) Bioactivation of low water-soluble-P phosphate rocks by phosphate-solubilizing bacteria. *In* Nutrient management for sustainable food production in Asia. IMPHOS-AARD/ CSAR Int. Conf., Bali, 9–12 Dec. 1996. Central Soil Res. Inst., Bogor, Indonesia, pp.68.

Goenadi DH, Siswanto and Sugiarto Y (2000) Bioactivation of poorly soluble phosphate rocks with a phosphorus-solubilizing fungus. *Soil Sci Soc Am J*, 64:927-932.

Goldstein AH (1995) Recent progress in understanding the molecular genetics and biochemistry of calcium phophate solubilization by gram negative bacteria. *Biol Agric Hort*, 12:185-193.

Guar AD, Ostwal KP and Mathur RS (1980) Save super phosphate by using phoshobacteria. *Kheti*, 32: 23-25.

Gupta RP, Singh P and Pandher MS (1998) Role of Phosphorus Solabilizing Microorganisms in P-economy and crop yield. *In : Soil Plant Microbe Interaction in relation to Integrated Nutrient Management*, Kaushik BD (ed), Division of Microbiology, IARI, ND, pp. 95-101.

Gyaneshwar P, Kumar GN, Parekh LJ and Poole PS (2002) Role of soil microorganisms in improving P nutrition of plants. *Plant Soil*, 245 : 83-93.

He ZL, Bian W and Zhu J (2002) Screening and identification of microorganisms capable of utilizing phosphate adsorbed by goethite. *Commun Soil Sci Plant Anal*, 33(5/6): 647–663.

Holford I (1989) Phosphate behaviour in soils. *Agricultural Sciences,* 2(5):15-21.

Igual JM, Valverde A, Cervantes E and Velazoez E (2001) Phosphate-solubilizing bacteria as inoculants for agriculture: use of updated molecular techniques in their study. *Agronomie,* 21:561–568.

Isherwood KF (2000) *Mineral fertilizer use and the environment.* International Fertilizer Industry Association / United Nations Environment Programme, Paris.

Jakasaniya MS and Trivedi BS (2004) Transformation of added phosphorus into inorganic P-fractions in soils of Gujarat. *Madras Agric J,* 91:163-167.

Johri JK, Surange S and Nautiyal CS (1999) Occurrence of salt, pH, and temperature-tolerant, phosphate-solubilizing bacteria in alkaline soils. *Curr Microbiol,* 39(2):89-93.

Kang SC, Ha CG, Lee TG and Mahehwari DK (2002) Solubilization of insoluble inorganic phosphates by a soil-inhabiting fungus *Fomitopsis* sp. PS 102. *Curr Sci,* 82(4): 439-442.

Kucey RMN (1987) Increased phosphorus uptake by wheat and beans inoculated with a phosphorus solubilizing *Penicillium bilaji* strain and with vesicular arbuscular mycorrhizal fungi. *Appl Environ Microbiol,* 52: 2699-2703.

Kucey RMN, Janzen HH and Leggett ME (1989) Microbially mediated increases in plant-available phosphorus. *Advances Agronomy,* 42, 199-228.

Kundu BS, Gera R, Sharma N, Bhatia A and Sharma R (2002) Host specificity of phosphate solubilizing bacteria. *Ind J Microbiol,* 42:19-21.

Macklon AES, Grayston SJ, Shand CA, Sim A, Sellars S and Ord BG (1997) Uptake and transport of phosphorus by *Agrostis inoculation* seedlings from rapidly hydrolysed organic sources extracted from ^{32}P-labelled bacterial cultures. *Plant Soil,* 190:163-167.

Mehana TA and Wahid OAA (2002) Associative effect of phosphate dissolving fungi, *Rhizobium* and phosphate fertilizer on some soil properties, yield components and the phosphorus and nitrogen concentration and uptake by *Vicia faba* L. under field conditions. *Pakistan J Biol Sci,* 5(11):1226-1231.

Mehta S and Nautiyal CS (2001) An efficient method for qualitative screening of phosphate-solubilizing bacteria.*Curr Microbiol,* 43(1):51-56.

Mikanova O and Novakova J (2002) Evaluation of P solubilizing activity of soil microorganisms and its sensitivity to soluble phosphate. *Rostlinna Vyroba,* 48(9):397-400.

Mishustin EN and Naumova AN (1962) Bacterial fertilizers, their effectiveness and mode of action. *Mikrobiologia,* 31:543-555.

Mueller T, Jensen LS, Magid J and Nielsen NE (1997) Temporal variation of C and N turnover in soil after oilseed rape straw incorporation in the field-simulations with the soil-plant- atmosphere model DAISY. *Ecol Modelling,* 99:247-262.

Oberson A, Friesen DK, Rao IM, Bühler S and Frossard E (2001) Phosphorus transformations in an oxisol under contrasting land-use systems: The role of the microbial biomass. *Plant Soil,* 237:197-210.

Orlova AI (2003) Obtaining and crystallochemical property of phosphates. *Nuclear chemistry,* 45(2):97-102.

Pikovskaya RI (1948) Mobilization of phosphors in soil in connection with the vital activity of some microbial species. *Microbiologiya,* 17:362-370.

Plaxton WC (2004) Plant response to stress: biochemical adaptations to phosphate deficiency. *Encyclopedia Pl Crop Sci.,* pp. 976-980, Online citation:DOI:10.1081/E-EPCS 120010648.

Prabhakaran J, Ravi KB and Srinivasan K (1996) Response of Vamban-1 blackgram to N_2 fixer and P mobilizers in acid soil. *In: Microbiology Abstracts, XXXVII Annual Conference of the Association of Microbiologists of India.* IIT, Chennai, Dec4-6, 1996, p. 120.

Radhakrishnan KC (1996) Role of biofertilizers in cotton productivity. Paper presented in National Seminar on Biofertilizer Production-Problem and Constraints, TNAU, Coimbatore, Jan 24-25, 1996, p.17.

Raja AR, Shah KH, Aslam M and Memon MY (2002) Response of phosphobacterial and mycorrhizal inoculation in wheat. *Asian J Plant Sciences,* 1(4): 322-323.

Raju RA and Reddy MN (1999) Effect of rock phosphate amended with phosphate solubilizing bacteria and farmyard manure in wetland rice (*Oryza sativa*). *Indian J Agril Sci*, 69:451-453.

Randall PJ, Hayes JE, Hocking PJ and Richardson AE (2001) Root exudates in phosphorus acquisition by plants. *In: Plant nutrient acquisition: New perspectives* (Eds N. Ae, J. Arihara, K. Okada and A. Srinivasan), pp.71-100, Springer-Verlag, Tokyo.

Ranjard L, Poly F and Nazaret S (2000) Monitoring complex bacterial communities using culture-independent molecular techniques: Application to soil environment. *Research Microbiol*, 151:167-177.

Richardson AE (1994) Soil microorganisms and phosphorus availability. *In* Soil biota: Management in sustainable farming systems, Vol. 1 (Eds, Pankhurst, C. E., Doube, B. M., Gupta, V. and Grace, P. R.) CSIRO Publishing, Adelaide.

Schilling G, Gransee A, Deubel A, Lezovic G and Ruppel S (1998) Phosphorus availability, root exudates, and microbial activity in the rhizosphere. *Z Pflanzenernähr Bodenk*, 161:465-478.

Subba Rao NS (1982) Biofetilizer in Agriculture. New Delhi, India, Oxford and IBH Publishing Co.

Subba Rao NS (1995) Biofertilizers in Agriculture and Forestry, Oxford and IBH Publishing Co., New Dehi, pp. 3.

Sundara Rao WVB (1968) Phosphorus solubilization by microorganisms. *In: Proceedings of All Indian Symposium on Agricultural Microbiology*, University of Agricultural Sciences, Banglore, India, 210-229.

Tarafdar JC and Marschner H (1995) Dual inoculation with *Aspergillus fumigatus* and *Glomus mossae* enhances biomass production and nutrient uptake in wheat (*Triticum aestivum* L.) supplied with organic phosphorus as Na-phytate. *Plant Soil*, 173:97-102.

Theron J and Cloete TE (2000) Molecular techniques for determining microbial diversity and community structure in natural environments. *Crit Rev Microbiol*, 26:37-57.

Thien SJ and Myers R (1991) Determination of bioavailable phosphorus in soil. *Soil Sci Society America*, 56:814-18.

Van Breemen N, Finlay R, Lundström U, Jongmans A, Giesler R and Olsson M (2000) Mycorrhizal weathering: A true case of mineral plant nutrition? *Biogeochemistry*, 49:53-67.

Whitelaw M (2000) Growth promotion of plants inoculated with phosphate-solubilizing fungi. *Advances Agronomy*, 69:99-151.

Mind Bogglers

1. Why available soil phosphorus remains low even after input of chemical forms of P (inorganic P fertilizers)?
2. What are the probable mechanisms of P solubilization by virtue of activity of microbe/microbial enzymes?
3. Enlist the genes involved in P-solubilization if the mechanism of P-solubilization is via production of organic acids.
4. Suggest some possible reasons for the inconsistency of performance of PSB/PSF(s) under field conditions.
5. How the competitiveness of inoculated P-solubilizers could be enhanced for better establishment and survival in the rhizospheric soil or rhizoplane region?

❑❑❑

Appendix

List of genera of microbes involved in solubilization of inorganic phosphates in soil

Bacteria	**Fungi**	**Cyanobacteria**
Achromobacter sp	*Alternaria* sp	*Anabaena* sp
Agrobacterium sp	*Aspergillus* sp	*Calothrix brauni*
Alcaligenes sp	*A. awamori*	*Nostoc* sp
Azotobacter chroococcum	*A. candidus*	*Scytonema* sp
Aerobacter sp	*A. carbonum*	*Tolypothrix ceylonica*
Arthrobacter sp	*A. flavus*	
Bacillus sp	*A. foetidus*	Actinomycetes
B. cereus	*A. fumigatus*	*Micromonospora* sp
B. firmus	*A. nidulans*	*Nocardia*
B. megaterium	*A. niger*	*Streptomyces* sp
B. mycoides	*A. terreus*	
B. subtilis	*A. wentii*	
B. japonicum	*Cephalosporium* sp	
B. polymyxa	*Chaetomium* sp	
B. circulans	*C. globosum*	
B. licheniformis	*Cladosporium* sp	
B. mesentricus	*C. sphaerospermum*	
B. pumilis	*Cuvrularia* sp	
B. pulvifaciens	*C. lunata*	
Bradyrhizobium sp	*Cyatus africanus*	
B. japonicum	*C. bulleri*	
Brevibacterium sp	*Cylindrocarpon otusisporum*	
Citrobacter freundii	*Fusarium* sp	
Corynebacterium sp	*Fomitopsis* sp	
Enterobacter aerogenes	*Helminthosporium* sp	
Erwinia sp	*Humicola insolens*	
Erwinia freundi	*H. lanuginose*	
Escherichia sp	*Mucor mihii*	
Escherichia intermedia	*Myrothecium roridum*	
Flavobacterium sp	*Oidiodendron* sp	
Micrococcus sp	*Paecilomyces fusisporus*	
Pseudomonas sp	*Penicillum* sp	

□□□□□...

P. aeruginosa	*P. bilaji*
P. putida	*P. digitatum*
P. straita	*P. funiculosum*
Rhizobium sp	*Rhizoctonia solani*
Sarcina sp	*Rhizopus* sp
Serratia phosphaticum	*Shwanniomyces occidentalis*
Synechocystis	*Scleratium rolfsii*
Xanthomonas sp	*Trichoderma* sp
	Trichoderma sp
	Trichurus spiralis
	Torula thermophila

6

Vesicular Arbuscular Mycorrhizae

- Introduction 140
- VAM Systematics 142
- Life Cycle of VAM Fungi 148
- Physiology of AM Fungi 155
- Role as Biofertilizer 155
- Mass Production: Problems and Prospects 163
- Conclusions 170
- References 171
- Mind Bogglers 180

INTRODUCTION

Vesicular arbuscular mycorrhizae are fungi that exhibit symbiosis with roots of higher plants. The term vesicular arbuscular is derived from characteristic structure i.e. 'branched hyphae or arbuscles' that are formed with in the root cortical cells and vesicles or 'swollen lipid storing organs' that occur with in or between them.

The AM symbiosis represents an ancient symbiosis as palentological studies revealed the presence of AM fungal hyphae and arbuscules in fossils of *Aglaophyton* isolated from *Rhynie chert* in the early Devonian era (Taylor *et al*, 1997). The existence of mycorrhizal fossils from very early epochs and its prevalence over all symbioses between plants and fungi indicate an important role of AM symbiosis in the evolution of terrestrial life on earth. The invasion of land by the ancestors of vascular plants may have been facilitated by the occurrence of a symbiotic association and the symbiosis is influencing the terrestrial ecosystems until present times (Pirozynski and Dalpe, 1989). About 80% of all terrestrial plant species form VAM symbiosis (Smith and Read, 1997) and approximately 95% of the world's present species of vascular plants belong to families that are characteristically mycorrhizal (Quilambo, 2000) including roots of angiosperms, gymnosperms and pteridophytes inhabiting diverse environmental regimes or wide range of habitats. Recently aquatic plants or waterlogged plants are also reported to harbor this symbiosis (Beck-Nielsen and Madsen, 2001). However certain crop plants belonging to Cruciferae, Brassicaceae, Chenopodiaceae, and Caryophyllaceae families do not form symbiosis with AM fungi (Barker *et al*, 1998). These fungi have caused smooth transition of aquatic to terrestrial plants, have helped in establishment and maintenance of terrestrial plant ecosystems (Selosse and Le Tacon, 1998) and may have a pivotal role in determining plant community structure. Alteration of AM fungal diversity modifies the rhizospheric potentialities and soil fertility.

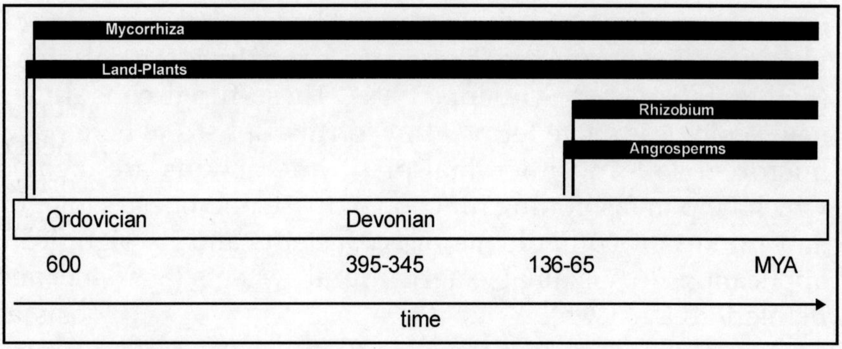

Origin of vesicular arbuscular mycorrhizae on the geological time scale

The AM fungi are obligate biotrophs, a feature unique to this class of microbes which demands the presence of the host/ host exudates and thus completion of their life cycle depends on their ability to colonize a host plant. This symbiosis is mutually beneficial since the fungus gets its share of plant's photoassimilates (in form of glucose taken up by the fungus at the plant-fungal interface) while the plant is benefited in form of increased supply of mineral nutrients provided by the fungus. Graham and Abbott (2000) have reported that increased AM colonization at moderate/high soil P content can reduce plant growth due to carbon costs (photosynthate recruited to AM fungi) without beneficial effects for the plant-nutritional status.

Symbiosis

- DeBary (1887) first introduced symbiosis
- also coined the distinction between mutualistic and antagonistic symbiosis
- described symbiosis to be association in which two partners benefit from each other
- antagonistic symbiosis, an association in which only one partner benefits from other

AM fungi not only form arbuscules but also develop extensive extraradical hyphal network extending out of the plant root system, which makes a significant contribution to the improvement of soil texture and water relations as soil particles bind on the surface of the hyphal cell wall (Bethlenfalvay and Schuepp, 1994). These fungi constitute a new ecological niche termed 'mycorrhizosphere' where novel microbial-soil and microbial-plant interactions are created which help in fashioning function of these microbes as integral and important components of ecosystems and its significant applications in sustainable agricultural systems (Schreiner and Bethlenfalvay, 1995).

In soybean, mycorrhizal plants exhibit enhanced P absorption and increased P content and dual inoculation with nitrogen fixing *Rhizobium* and AM fungus *Glomus versiforme* further increased P-uptake (Chang *et al*, 1992). However, AM symbiosis in plants is affected by a range of agricultural practices such as fertilizer application, crop rotation, tillage, and liming. These affect potential of AM under field conditions and root colonization levels (Dalpe and Monreal, 2004). The dependency of plants for undergoing AM association varies as plant species differ in their P requirements for example beans, corn and leek have a much higher mycorrhizal dependency than potato and wheat. The AM symbiosis shares similarities and common pathways with the *Rhizobium*-legume interaction. However this old symbiotic association has remained obscured in terms of molecules engaged in signaling processes and other biological/genetic aspects for several decades due to obligate status, coenocytic nature of AM spores and recombination.

VAM Systematics

Early work by Peyronel (1923) established the regular association of very large spores and sporocarps of members of this family with AM roots. Later, Butler (1939) and Mosse (1956) described them as the chief causal organisms of AM infections.

AM systematics has been an ever evolving limb of science. These fungi are the members of class Zygomycota and according to the current classification system this class contains one order, the Glomales, encompassing six genera and approximately 150 species (Morton *et al*, 1993). AM systematics was revised according to a phylogenetic perspective (Morton and Benny, 1990), which separated a new order 'Glomales' from Endogonales'. The order Glomales now contains two suborders: Glomineae and Gigasporineae'. All members of both suborders contain aseptate, multinucleate vegetative mycelium that forms intraradical structures in host roots. More recently Redecker *et al* (2000) combined estimates of fungal phylogeny derived from ribosomal small subunit sequences and fossil records to conclude that the major groups of terrestrial fungi originated around 600 million years ago.

Morphological criteria

AM fungi belong to new phylum of Glomeromycota, class Glomeromycetes and one order Glomales, divided to two suborders, three families and six genera of which around 150 species have been described (Helgason *et al*, 1999). The broad host range and the ability to spread easily in nature helped these fungi (may be small number of species) to colonize up to 80 per cent of the land plant species rather the origin and evolution of land plants is linked to these fungi.

These fungi have been classified on the basis of morphological traits of the spores (the most frequently used criteria) and other fungal structures (color, size, morphology, cell-wall structures). These criteria help to differentiate between various AM fungal species and provide information for phylogeny of the order Glomales. This order is further classified into two suborders *viz Glomineae* and *Gigasporineae,* the latter has the family *Gigasporaceae*, whose members exhibit abundant coiled, swollen hyphae and arbuscules and produce extraradical auxiliary cells.

History of AM fungal research

- Nageli (1842) first description of arbuscular mycorrhiza
- early observations of symbiosis Trappe and Berch (1985)/Rayner (1926–1927)

- basic anatomical relationships between host and fungal tissues observed by Schlicht (1889)
- Janse (1897) termed intramatrical spores as 'vesicles'
- Gallaud (1905) named "arbuscules" (structures located in the inner cortex)
- partial digestion of the arbuscule result in a sporangiole/vesicle Gallaud (1905)
- Electron microscopic (TEM) studies to confirm intracellular nature of arbuscules and sporangiole by Cox and Sanders (1974)
- Types of arbuscules (Arum and Paris) by Gallaud (1905)
- Jones (1924) described appresorium formation
- light and electron micrographs to ascertain AM fungal internal structures Bonfante (1991)
- Scannerini and Bellando (1968) space between host membrane and AM wall contain materials of host origin
- Phillips and Hayman (1970) gave differential lactophenol trypan blue staining technique
- Newman (1966) devised line intersect technique for quantification of mycorrhization
- Magrou (1946) first observed fungal growth in hanging drop cultures in pieces of surface sterilized potato roots
- Nicholls (1952) first isolated strain *Pythium ultimum* from mycorrhizal onion roots
- Mosse (1961) developed *in vivo* cultivation technique in pots
- Mosse (1953) first performed successful "AM infection" of strawberry by *Glomus mosseae*.
- Gerdemann (1955) isolated *Gigaspora gigantean*

Koide and Moose (2004)

The family *Gigasporaceae* is comprised of two genera *Gigaspora* and *Scutellospora* that are differentiated by absence and presence, respectively, of hyaline flexible inner spore walls (Morton and Benny, 1990).

The suborder *Glomineae* has two families, *Glomaceae* and *Acaulasporaceae,* in which the fungal species produces arbuscules as well as vesicles but no auxiliary cells. These two families contain two genera respectively, *Glomus* and *Sclerocystis* for *Glomaceae* and *Acaulospora* and *Entrophospora* for *Acaulasporaceae*. The *Acaulosporaceae* and *Gigasporaceae* families were supposed to have diverged from each other around 250 MYA ago (Simon *et al*, 1993) while the *Glomus* species is speculated to have appeared first and the two families later in the due course of evolution.

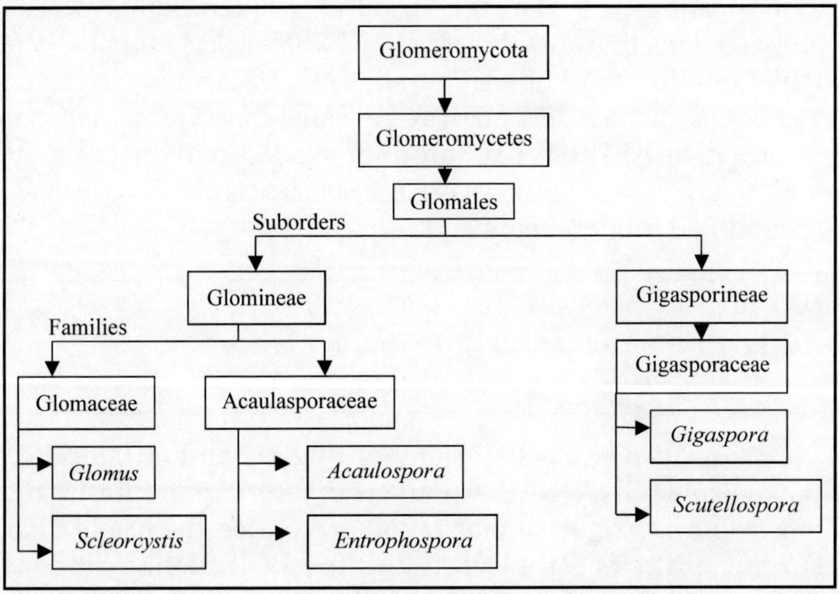

Flow diagram showing the taxonomic status of the six well-known genera of VAM fungi

Molecular approaches

The past decade has witnessed inrush of information on molecular genetics and biochemistry of VAM fungi despite peculiar problems in AM fungal study as inability to grow in pure culture. Molecular biology techniques have been very successful for genetic characterization of these microorganisms as these approaches provided reliable and useful information in order to identify and differentiate fungal isolates. Several molecular markers have been designed for deciphering the molecular genetics of AM fungi. The first DNA fungal regions that were sequenced and used as targets for phylogenetic analyses on the AM fungi were rRNA gene parts, including the large subunit (LSU), small subunit (SSU) and the internal transcribed spacers (ITS) (Redecker *et al*, 1997). The sequence information of SSU and ITS sequences could be coupled to design specific primers for identification of AM fungi from both spores and fungal structures within plant roots by PCR amplification (Harrison, 1999). rDNA sequence analysis also allowed detection and classification of novel AM fungal species that extended the knowledge about various soil ecosystems. The SSU sequence data analysis revealed the approximate date of origin of AM fungi i.e. around 462-353 million years ago and divergence of families as *Gigasporaceae* and *Acaulosporaceae* around 250 million years ago.

DNA content of the hyphal germ tube

- range from 0.13 to 10 pg per nucleus
- varies from one species to other

Lanfranco *et al* (1995) have utilized random amplified polymorphic DNA (RAPD) markers to ascertain specific isolates of *Glomus mossae*. A similar report advocated the use of taxon specific primers for ascertaining the PCR identification of different AM fungi co-existing in the same root fragment (van Tuinen *et al*, 1998).

Since AM fungi exhibit enormous population diversity, DNA markers based on PCR are promising tools due to requirement of very small amounts of DNA. PCR techniques including the study of molecular markers as highly variable minisatellites (Gadkar *et al*, 1997) and amplified fragment length polymorphism (AFLP) (Rosendahl and Taylor, 1997) reveal high genetic variation in AM fungi at both interspecific and intraspecific levels. The analysis of ITS regions could also reveal the variability in the genetic composition/ heterokaryotic nature with in and between multinucleate spores of a single fungal species. A recent report advocates the combination of morphological and molecular approaches for characterization of *Gigaspora* isolates from different geographical areas for better clarification of relationships among species (Lanfranco *et al*, 2001).

Different genera and species of Arbuscular Mycorrhizal fungi

Colums microaggregatum	*Scelerocystis microcarpa*
G.diaphanum	*S.trap*
G.geosporum	*S.castanea*
G.inverimaium	
G.claroideum	*Acaulospora laevis*
G.clarum	*A.trappei*
G.coronatum	*A.morrowiae*
G.etunicatum	*A.denticulata*
G.fasciculatum	
G.fragilistratum	*Gigaspora candida*
G.geosporum	*G.margarita*
G.lamellosum	*G. aff. Margarita*
G.luteum	*G.albida*
G.macrocarpum	*G.trap*
G.manihotis	*G.rosea*
G.mosseae	
G.occultum	*Scutellospora gilmorei*
G.spurcum	*S.aurigloba*
G.versiforme	*S.pellucida*
G.viscosum	
Glomus caledonium	*Entrophospora infrequens*
	Geosiphon pyriforme

AM fungal diversity in India

Analysis of soil samples from various ecosites and geographical regions of country have demonstrated the occurrence of five types of VAM fungal genera with fourteen spore types of genera *Glomus viz Glomus macrocarpum, Glomus macrocarpum* Tul., *Glomus macrocarpum* Tul. var. *geosporum, G. mosseae, Glomus multicaule* (Gerdemann and Bakshi, 1976) and *Glomus* sp., *Sclerocystis viz Sclerocystis coremioides* Berk., *Sclerocystis coremioides* Broome, *Sclerocystis sinuosa* (Gerdemann and Bakshi, 1976), *Sclerocystis* sp., *Gigaspora viz. Gigaspora calospora, Acaulospora* sp and *Endogone viz Endogone gigantean* Nicol., *Endogone gigantean* Gerd., *Endogone microcarpum* and *Endogone sp* (Bakshi, 1974)

Later reports advocated the immense diversity of AM fungi in forest soils and soils of coastal area, sand dunes, arid zones, deserts, agricultural/crop fields, grasslands and stressed ecosystems as mines, heavy metal/petro/industrial/sewage/ tannery effluent polluted sites (Singh and Adholeya, 2002) from different parts including Delhi, Andhra Pradesh, Tamil Nadu, Karnataka, West Bengal, Goa and Kerala (Khade and Rodrigues, 2003; Manocharachary *et al*, 2005).

LIFE CYCLE OF VAM FUNGI

There is limited knowledge about physiology and complete life cycle of VAM fungi; still considerable work has been done on development and establishment of VAM symbiosis in host plant roots. As *Rhizobium*-legume interaction, this symbiosis also involves exchange of diverse signals between two symbionts, although the nature and action mechanism of these molecules remains unexplored. Host root exudates, especially (iso) flavonoids,enhance spore germination, and elongation and branching of hyphae (Giovannetti *et al*, 1993). There are several stages of infection and establishment such as:-

Pre-Infection/Early Signaling Events: The fungal chlamydospores that are present in soil get germinated to form the initial hyphal germ tubes even in absence of plant root but root exduates and volatiles as CO_2 stimulate both of these processes. The extended hyphal germ tubes undergo rapid and extensive branching on encountering host plant roots which suggests a recognition event governed by certain flavonoids or phenolic compounds or specific positive reaction in one fungus but inhibition in other.

Primary Infection/Appresorium formation: This stage involves differentiation of a fungal hypha to form an appressorium upon contact with root of host plant (no growth on contact with non-host root) and further penetration into the root between epidermal cells. Root exduates though stimulate hyphal growth and branching but appressoria are formed only on intact roots suggesting a particular signaling event. Fungus recognizes specifically epidermal cell wall, as appressorium is not formed on cortical or vascular cell walls. (Smith and Bowen, 1979). This reveals role of epidermal cell wall constituents including cellulose, polygalacturonan and epidermal wall proteins in signaling appressorium formation within epidermal cells. Out of these signaling molecules carbohydrates are more likely candidates for appressorium induction.

Penetration of Root: Appressorium formation is followed by development of a penetration hypha that goes into the root. Penetration may occur:-

a) Intercellularly; where hypha enters by forcing it's way between two epidermal cells.
b) Intracellularly; where hypha penetrates an epidermal or root hair cell wall and grows through cell.

The exact mechanisms involved in penetration are unknown but it has been suggested that specific, localized production of cell wall degrading enzymes in combination with mechanical force may facilitate entry of hyphae without inducing defense responses. Production of xyloglucanases and polygalacturonases expedite AM hyphal passage through a cell wall. The root penetration event is plant controlled eg. cytochemical studies of mutant *Pisum sativum* and *Medicago sativa* exhibited appressoria formation and presence of phenolics adjacent to appressoria, which suggests elicitation of a defense response though no penetration.

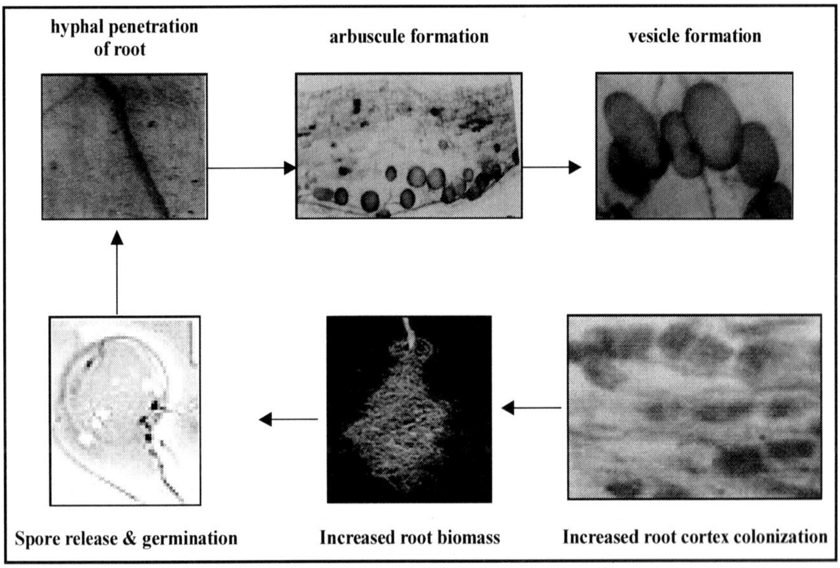

Propagation cycle of vesicular arbuscular mycorrhizal (VAM) fungi

(For a coloured version of this figure, see plate section, page 301)

Internal Growth and Development of Characteristic Structures

The plant influences internal development of fungus and a single fungal species may show significantly different morphological growth patterns depending upon the plant partner in association. Broadly there are two patterns of AM internal growth and development *viz*. Arum and Paris type. The former one involves initial root penetration event followed by intercellular hyphal growth. On reaching inner cortex, branches arising from intercellular hyphae penetrate the cortical cell walls and differentiate terminally with in the cell to form dichotomously branched tree-like structures called 'Arbuscules'. Arbuscule remain apoplastic as plant plasma membrane extends to completely surround it. Arbuscule development is accompanied by few structural changes in the fungal cell wall which progressively becomes thinner giving extensive intracellular interface and works as a part of symbiotic interface or site for transport of nutrients and minerals between both symbiont (Peterson and Farquhar, 1994).

Arbuscules are short lived and hence remain functional for few days and then get collapsed and decay leaving the cell undamaged and capable of hosting another arbuscule. Arbuscule formation may be followed by lipid filled storage organ, vesicle formation with in roots in some species of AM fungi.

Phase contrast micrographs of arbuscular structures formed intracellularly in host

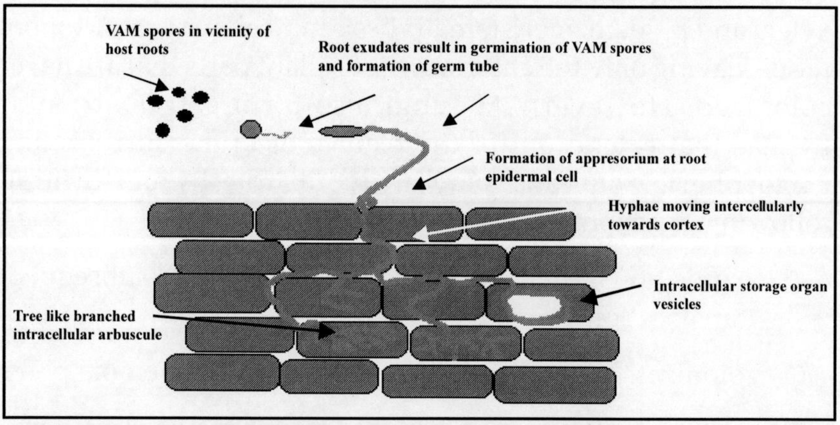

Diagrommatic longitudinal section showing formation of germ tube, appresorium, characteristics arbuscules and vesicles by VAM fungi in host root cells *(For a coloured version of this figure, see plate section, page 302)*

Cellular and molecular alterations during arbuscule development

As the hypha penetrates the cortical cell wall and begins to differentiate into an arbuscule, the cell wall being invaded responds with:-

- fragmentation of vacuole
- migration of nucleus to a central position within cell
- an increase in number of organelles

Arbuscular interface development in cortical cells is accompanied by several alterations:

- Transcription of auxin inducible gene encoding the glutathione-s-transferase (GST) is induced
- GSTs are stress-response enzymes
- add Glutathione moiety to phenylpropanoids/flavanoids and other compounds
- glutathione moiety addition to phenolics facilitates their transport to the vacuole via specific carrier
- this sequestration prevents cytotoxic/phytotoxic effects of these compounds on cells at elevated levels

Fungal Extension

The fungal growth occurs in the epidermis, endodermis, xylem and phloem, meristematic tissue and above ground shoot tissue leaving only the chlorophyllous plant parts that are never colonized. However AM fungi also form extramatricular hyphae that extend out of the plant root in the surrounding rhizospheric zone and play a pivotal role by performing following functions:

- acquisition of mineral nutrients from soil and subsequent translocation to infected host plant
- colonization of additional roots and production of spores
- contributes to soil stability by aggregation of soil particles, probably mediated by glycoproteins (in part) produced by the hyphae.

Reproductive Structures

New spores are formed and are released into soil in the vicinity of roots. These chlamydospores under unfavorable conditions serve as means of survival. The spores have very thick and hard coating and may also have a hooked hinge like structure. New vegetative cells and later hyphae germinate out of the spores on advent of favourable conditions.

Genes Induced during AM Symbiosis

As discussed in the earlier sections of the chapter that AM fungal association result in formation of peculiar structures inside the host plant roots and the infection events are signal controlled. AM fungi share certain similarities with rhizobial-legume association as few reports suggest the presence of common receptors for both rhizobia and AM fungi though the morphological responses that occur in epidermal and cortical cells for both of these symbioses appear to involve unrelated processes at superficial context. The DNA sequence studies have provided striking similarity among infection processes of these symbioses at the molecular genetic level. Several genes have been identified which are induced during both symbiotic interactions,e.g. the early nodulin genes *ENOD2*, *ENOD40* (Van Rhijn *et al*, 1997). However, the most convincing evidence that the infection processes used by both microsymbionts (rhizobia and AM fungi) involve common steps came from studies with legume mutants, which have lost the ability to form nodules. A large proportion of the nodulation-resistant mutants were also completely resistant to AM fungi, while their interaction with soil pathogens was not affected (Harrison, 1997).

Initial comparisons of protein profiles of colonized and non-colonized roots showed presence of new proteins in mycorrhizal roots but provided differentiation between fungal and newly induced plant proteins.

- ❒ Molecular analysis of cDNA and genomic sequence comparisons revealed that intrinsic membrane protein

(IMP) synthesis is induced in mycorrhizal roots. These IMPs play an important role in transport at arbuscular interface.

- Plant Cell-arbuscular interface require denovo synthesis of cell wall proteins hence genes encoding HRGP (hydroxyproline rich glycoprotein) and a member of XET family (xyloglucan endo-transglycosylase) are induced during development of association. XETs cleave and reform xyloglucan bonds in cell wall and thus loosen cell walls to permit penetration of fungal hyphal or maintain structure of arbuscular interface matrix
- Biosynthesis of membrane is increased having high ATPase activity required for active transport of nutrients across the membrane
- Increased synthesis and deposition of β-1,4 glucans
- Increased synthesis of cytoskeleton components (activation of matrix α-tubulin promoter specifically in cells containing arbuscules) required for significant reorganization of cytoskeleton of cell during formation of arbuscular interface
- Down-regulation of certain genes, e.g. down regulation of four phosphate starvation inducible genes in *Medicago truncatula*, down-regulation of defense related genes such as genes coding for chitinases, glucanases or pathogenesis related proteins

Saif (1977) and later Douds and Chaney (1982) summarized the pattern of mycorrhizal infection within roots in three phases viz;

- *Lag Phase:* gradual colonization of root system by mycorrhizal fungi. The duration of this phase is related to the time required for spore germination, germ tube growth and host penetration by the fungus
- *Phase of Extensive Mycorrhizal Development:* formation of extensive mycelium in soil, initiating new colonies in the roots
- *Constant Phase:* colonization proportion of mycorrhizal to non-mycorrhizal roots rarely exceeds 80 per cent. There also occur physiological changes in host roots and host related to reproductive growth of fungus

Physiology of AM Fungi

The physiological aspects of AM fungi involve mechanisms that account for the mycorrhizal effects on plant growth and nutrient. This effect prophecies the potential of VAM as a biofertilizer and bioprotectant for enhancing plant growth and yield, as soil conditioner and as a prime constituent of bioremediation brigade for sustainability of the rapidly degrading environment (Gaur and Adholeya, 2004). The fungal association enriches plant nutrient in lieu of which the former is nourished by plant host by offerings in form of photosynthate/sugars/carbon source. This bi-directional transport leads to enhanced plant growth and completion of fungal life cycle (Pfeffer *et al,* 1999).

VAM symbiosis enhances host root's ability to absorb less mobile/soluble minerals as P, Cu and Zn from soil. This improved mineral nutrition enhances plant growth as well as plant biomass and also influences partitioning of this material between root and shoot.

Role as Biofertilizer

The AM colonized plants are better nourished and better adapted to their environment. AM fungal associations result in increased P nutrition and the plant cherishes other benefits of association as increased protection against environmental stresses (Gupta et al, 2007) including cold, salinity, drought, pollution (Shetty *et al,* 1995) and reduction in incidence of root diseases and harmful effects of certain pathogenic agents (St-Arnaud *et al,* 1995).

Increased P availability to plants

Phosphorus in soil occur as inorganic (P_i) or organic (P_O) phosphate pools. The proportion of total soil P in solution is usually less than 1 per cent. The inorganic P (P_i) is held firmly in crystal lattices as insoluble Ca^{2+}, Fe^{2+} and Al^{3+} phosphates constituting the unavailable P pool of total soil P which is not absorbed by the plants, while less tightly bound P_i get exchanged rapidly in soil solution and become readily available at pH 6.5 to plants by diffusion in root hence creating a depletion zone. On addition of P fertilizer, a short-term increase in PO_4^{2-} ions does occur which is later removed by fixation. The other kind of soil P pool includes the organic P (P_0) that occurs in form of a variety of phosphorylated compounds *viz* phosphatidyl inositol, phospholipids and nucleic acids and hydrolysis of such compounds yields P. The availability of this fraction of soil P to plants depends upon lysis by microorganisms or enzymes released by them.

Increase in the root absorptive surface area by extended extramatricular hyphae

The extent of P availability by AM symbiosis depends on the level of applied P fertilizer, high doses of P fertilizer slow down or inhibit mycorrhizal efficiency. Mycorrhizal roots are more efficient in nutrient uptake than non-colonized roots and hence plants exhibiting this symbiosis are recorded to be large with more P content in comparison to control plants. Increased P uptake result in growth improvement of plant, which may be attributed to enhanced mineral uptake due to vast increase in the absorptive surface area of plant. The combined surface area of the millions of hyphae is far greater than that of non-mycorrhizal roots. Moreover, the extending hyphae are able to draw on more distant or inaccessible supplies of nutrients from regions (beyond the depletion zone) than the roots can reach. It is estimated that nearly 25-30% of phoshatic fertilizer can be saved through inoculation with efficient VAM fungi and yield is also improved. Jones *et al* (1998) examined P uptake in *Eucalyptus coccifera* seedlings, which can form both EM and AM fungal association and observed mycorrhizal interaction resulted in at least a 40% increase in P concentration in host plant shoots to that of uninfected plants. The AM fungi may account for as much as 80% of the P absorbed by the mycorrhizal roots and the transfer of P to plant is always in its inorganic phosphatic form.

Mycorrhizal systems depend on the ability of fungal symbionts to absorb nutrients that are available in organic or inorganic form and to translocate it or it's metabolites to plants through extensive vegetative mycelium. The extramatricular hyphae perform nutrient uptake by absorption of nutrients from soil solution followed by nutrient translocation in soil and into intramatricular fungal structures inside roots and finally nutrient transfer from fungus to plant across enlarged symbiotic interface at the arbuscules. A good correlation between P uptake and colonization has been found in *Hyacinthoides* (Merryweather and Fitter, 1996). The absorbing tissue involved in uptake is most difficult to determine and expressed on basis of root length defined as 'inflow' ($molm^{-1}s^{-1}$) by Brewster and Tinker (1972) or on basis of root weight to give specific absorption rate in $molm^{-1}s^{-1}$ (Smith and Gianinazzi-Pearson, 1990).

Phosphorus uptake by mycorrhizal rootlets from soil is an active process carried on by utilization of energy currency of cell produced during metabolic processes and hence is observed to reduce by using respiratory or metabolic poisons/inhibitors. Phosphorus is probably absorbed as $H_2PO^-_4$ via high or low affinity uptake systems. A PO^-_4 transporter (GvPT) has been cloned from *Glomus versiforme* (Harrison and Van Burren,

1996). This c-DNA gene is expressed in external hyphal of mycorrhizal roots and when transformed into mutant yeast (Pho 84) lacking PO^-_4 transporter, restared wild-type transport properties. Uptake is likely to be via proton co-transport with necessary PMF generated by a membrane-bound H^+ATPase as indirated by as HTATPARE inhibitor compound diethylstilbestroll applied resulted in unavailability of nutrient. More recently, a PCR cloning approach based on the use of highly degenerate primers allowed the isolation of five partial genomic clones encoding plasma membrane H^+-ATPases from the AM fungus *Glomus mosseae* (Ferrol *et al*, 2000). Ruiz-Lozano *et al* (2002) constructed a cDNA library of RNA from *Glomus intraradices*-colonized lettuce roots to identify gene of fungal origin cDNA (*Gi-1*), expressed only in mycorrhizal roots that exhibited little sequence homology with glutamine synthetase from *Helicobacter pylori* of ORF encoding 157 amino acids protein.

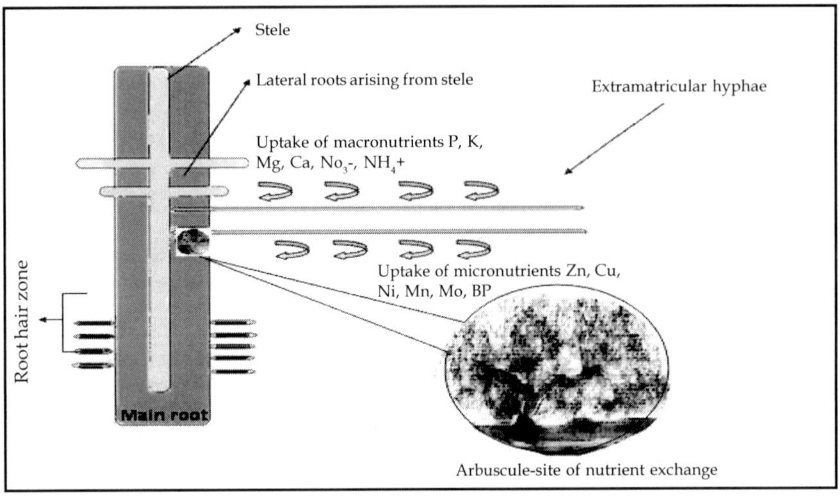

Schematic representation of uptake of mineral ions from soil by root in presence of AM fungi

Fungal cells synthesize polyphosphates and store/sequester them in their vacuoles or cytoplasm following uptake which is characteristically revealed by staining with toluidine blue stain. The alkaline phosphatses (enzymes) can later break the stored polyphosphates and transfer it to the host plant cell by cytoplasmic streaming of arbuscules via symbiotic interfaces.

Effect on nodulation and nitrogen fixation

In general mycorrhizal plants exhibit increased rate of photosynthesis and thus AM association affects plant growth however activities of other enzymes or enzyme systems can also be directly affected by P supply such as nitrate reductase and glutamine synthetase as well as N fixation in both legumes and non-legumes. AM infected plants exhibit increased N concentrations and this indicates one generalization that the fixed dinitrogen in rhizobial roots is assimilated well which result in increased N content under low phosphate soil conditions (Asai, 1944). Mycorrhizal plants show improved nodulation and nitrogen fixation may be due to relief from P stress and due to uptake of some essential micronutrients, which result in general improvement in growth and pose indirect effects upon the N_2-fixing system. Kaldrof *et al* (1994) reported presence of assimilatory nitrate reductase (NR) gene localized in only arbuscules (differentially functioning depending on different symbiotic stages), predicted to be of importance in N uptake from spores of *Glomus* using a PCR based strategy.

Improved P nutrition as well as direct fungal effects may be implicated in enhanced uptake by plants of other macronutrients as potassium and sulphate and of micronutrients as copper, zinc etc. Mycorrhizal plants accumulate lower amounts of organic anions in their vacuoles and thus may regulate cytoplasmic pH by disposing of a higher proportion of OH^- generated during nitrate reduction to the soil and might also represent saving in field carbon.

Tolerance to abiotic stresses

Heavy metal tolerance: The modern world has witnessed dramatic increase in heavy metal pollution of the soil ecosystems due to several human activities possibly including mobilization of these metals up in the food chain thereby threatening human health (Rabie, 2005b). Certain essential microelements are required in traces and accumulation of such elements at high concentrations in plants leads to toxicity and necrosis/death of plant. Thus, heavy metal toxicity may derive from excessive uptake of Zn^{2+}, Cu^{2+}, Fe^{2+}, and Co^{2+} as well as from other elements and ions which are normally regarded as toxic as Pb, Cd^{2+}, Ni^{2+}, Ti, Ba, etc. Heavy metal toxicity may result from binding of metals to sulphydral groups in proteins leading to inhibition of activity or disruption of structure or displacement of an essential element resulting in deficiency effects (Hall, 2002). Moreover excess of heavy metals stimulate formation of free radicals and reactive oxygen species, which may cause oxidative stress (Dietz *et al,* 1999) while other processes of toxicity include oxidation and cross linking of protein thiols, inhibition of key membrane proteins (H^+ ATPase), changes in composition and fluidity of membrane lipids.

AM fungi increase shoot uptake of metals in slightly metal contaminated soils (Weissenhorn *et al,* 1995), while in severely contaminated soil, could reduce shoot metal concentration and protect plants against harmful effects of metals (Malcova *et al,* 2003). Moreover at low soil pH arbuscular mycorrhizal symbiosis result in increased accumulation of Cu, Ni, Pb and Zn in grass *Ehrhartia calycina* (Killhan and Firestone, 1983), however, AM fungal symbiosis results in increased heavy metal tolerance at certain specific soil pH values (El-Kherbawy *et al,* 1989). Rabie (2005a) reported a high proportion of Zn, Cu, Pb and Cd in red kidney and wheat plant tissues that grew at heavy metal contaminated soil on account of mycorrhizal acquisition. The higher heavy metal concentration in AM plants could be explained by the fact that AM infection increased plant uptake of metals by mechanisms such as enlargement of the absorbing area, volume of accessible soil, binding capability for metals of the relatively large fungal biomass associated with the host plant roots and efficient hyphal translocation (Yu *et al,* 2004).

Arsenate tolerance in plants is most striking as it appears to depend on modifications to the P transport system and hence the root absorbing capacity (Mehrag, 1994). Tolerant genotypes, lead to the idea that tolerant plants are dependent on fungi for P uptake and consequent success (Mehrag *et al*, 1994). Physiologically, dilution of a toxic element can occur as a consequence of improved P nutrient and increased plant growth, even in situation where uptake of heavy metal per plant is actually greatly increased. The introduction of an AM fungal inoculum into heavy metal contaminated sites/areas could be a strategy for enhancing the establishment of mycorrhizal herbaceous species and help in revegetation of metal-contaminated mine spoil or other degraded sites (Gaur and Adholeya, 2004) though AM fungal isolates differ in their effect on heavy metal uptake by plants.

Drought tolerance and water relations: Several eco-physiological studies have demonstrated that AM symbiosis often results in altered rates of water movement into, through and out of the host plants, with consequent effects on tissue hydration and plant physiology. It is now accepted that the contribution of AM symbiosis to plant drought tolerance is the result of accumulative physical, nutritional, physiological and cellular effects (Ruiz-Lozano, 2003). There occur alterations in water relations of plants as a result of mycorrhizal infection. Mycorrhizal plants may have improved drought tolerance but it is difficult to distinguish direct mycorrhizal effects from those that could be mediated via improved mineral nutrition. Broadly, mycorrhizal plants have higher hydraulic conductivity especially when P availability is low (Auge, 2004). Shoot resistances are frequently reduced, with the main effect on stomata so that both CO_2 and H_2O movement are affected and rates of transpiration and photosynthesis are increased.

VAM colonization can reduce hydraulic resistance to water uptake in roots may be by increase in size or the branching of root system. Recent consensus says that hyphae exhibit conductance but it is difficult to deny that there are complexities in measurement and dependence of hydraulic conductivity on rate of transpiration and size and branching of root systems.

Mycorrhizal colonization may have effects on drought tolerance that aren't directly related to water relations. In dry soil, P and other nutrients (e.g NO_3^-) become less mobile and a mycorrhizal pathway of nutrient acquisition become relatively important. Relief of nutrient stress might also be followed by increased rates of root growth and more efficient extraction of water from the soil profile and nutrients, which are absorbed against the diffusion constant. Reduction in the severity of symptoms of drought stress such as proline accumulation follow this improved water availability. The non-mycorrhizal plants exhibit slow root growth that not only limits nutrient availability but also limits the plant access to water reserves in soil.

Quilambo (2000) reported that inoculation of peanut with an indigenous AM fungal inoculant resulted in increase in leaf and root growth and this also prevented expected increase in root to shoot ratio and root-weight ratio, normally observed under phosphorus deficient and drought stress conditions. A similar report by Kaya *et al* (2003) showed that mycorrhizal colonization improved plant yield and water use efficiency and also the quality of fruit in watermelon (*Citrullus lunatus* Thunb.).

Non-nutritional modifications of host physiology

Mycorhizal colonization and infection alters phytohormone accumulation in host tissues and there occurs changes in the levels of several plant hormones *viz* cytoknin, abscisic acid and gibberellin-like substances. Though the exact mechanism(s) of action of these alteration is not deciphered but various hypotheses correlate these changes to improved nutrient status of host while others link it with the fungal production of these hormones however it seems unlikely that phytohormone synthesis by fungus could account for the magnitude of the increase observed in mycorrhizal plants.

AM fungi and several biocontrol agents as *Burkholderia cepacia, Pseudomonas fluorescens, Trichoderma harzianum,*

Verticillium chlamydosporium exhibit a kind of synergism among themselves which explains the pivotal role of AM fungi in soil microbial and plant interactions as these fungi might affect plant and soil microbial activity by stimulating the production of root exudates, phytoalexins, and phenolic compounds. Microbes functioning as biocontrol agents and plant growth promoting bacteria (PGPB) get a new kind of habitat in form of 'Mycorrhizosphere' in presence of extramatricular hyphae of AM fungi and this region ushers the most nourished hot spots for these microbes in soil (Duponnois *et al,* 2005). Increase in synthesis of secondary metabolites such as lignin, ethylene and phenols as well as phytoalexins may all contribute to these protective effects.

Mass Production: Problems and Prospects

AM fungi possess several properties that iconize these microbes to be considered and applied as basic facet of sustainable agriculture owing to numerous reports of the benefits of its inoculation in crop plants and tree species. Inspite of economical and ecological benefits attached to AM use their large-scale production is the limitation that haunts its wide spread popularity. This obstacle may be ruled out by enriching our understanding of the genetic control of fungal growth and physiological integration between symbionts but a little rapid methods are required to obtain temporary solutions for mass cultivation and production of AM inoculum.

Being obligate symbionts AM fungi could be mass-produced only in the presence of living roots with presence of a host plant and growth medium becomes prerequisite for inoculum production. The most widely used and economical way is using soil as matrix, which provides optimum conditions for growth and reproduction of the fungus. Since AM fungal associations are universal and have been reported in almost all terrestrial plants, these can be reproduced on a wide range of host plants. AM fungal inoculum must promote economically significant growth or stress tolerance and the inoculum must be free of pathogens and contaminants to increase the efficacy and shelf life of the product. There are several possible techniques reported for mass production of AM inoculum and include

a) *in vivo culture*

Usually, AM fungi are grown on roots of greenhouse plants and chopped mycorrhizal roots, often mixed with the growth media containing hyphae and spores, are used as source of inoculum. To avoid contamination, the spores collected from soil samples can be surface sterilized with antibiotics (Mertz *et al*, 1979). Soil could be replaced by inert substances such as vermiculite, perlite, expanded clay aggregates, sand or a mixture of these for crude inoculum production.

The *in vivo* method of AM mass culture production is more beneficial to the farmers if organized medium farm owners to large farm owners could generate AM inocula at their own land. On farm production of AM inoculum is the cheaper and economical way as this would save farmers from associated processing and shipping costs and in addition farmers could produce locally adapted isolates and generate a taxonomically diverse inoculum (Douds *et al*, 2005).

b) in vitro culture/axenic culture techniques

i) *Solution culture:* Though economical, the solid carrier based inocula have a constraint of being bulky as well as require substantial amount of efforts for maintaining the uncontaminated status of the product during/after packaging and dissemination. Solution culture or soil less culture is one such alternative for remediating these constraints of solid based inocula. The technique involves growing infected roots in aqueous medium enriched with mineral nutrients required for the growth of roots under controlled abiotic and biotic conditions that further minimizes the chances of contamination. Several modifications of the basic protocol or procedures have occurred over the years and now a days nutrient film technique

(NFT) has been most widely used. This involves bathing the roots of the colonized plants by a thin film/mist of nutrient solution under better-aerated conditions in contamination free chambers. The solid mat of roots produced by NFT grown plants seemed ideal for the production of an easily harvested and less bulky form of inoculum than that produced by plants grown in soil or other solid media. A hydroponic system also allowed the extension and maintenance of growth of AM fungi and the extramatrical mycelium may be produced in nutrient solution, free of plant tissue.

Inspite of a better edge in terms of less bulk and higher number of infective tissue certain drawbacks of technique limits its popular use *viz.*

- ❒ heavy contamination of the inoculum by microfauna including saprophytic cell worms (Mosse and Thompson, 1983).
- ❒ no suitable method for storing the inoculum in the form of roots so formed
- ❒ requires constant monitoring and adjustment of nutrient concentrations varying also with the stage of plant growth.
- ❒ highly skilled operation requiring sophisticated equipment (Mosse, 1990).

ii) *Aeroponic culture method:* This method involves applying a fine mist of nutrient solutions to colonized roots for AM fungal inoculum production. It has been shown to produce high levels of contamination free fungal spores as well as highly infective root fragments that remain infective for upto 23 months. Keeping the inherent problems to commercialization such as bulkiness of system and biotic and abiotic factors limiting growth of AM fungi, i*n vitro* culture techniques are found to be a good alternative for studying factors affecting AM infection as there is complete control of chemical and physical characters of growth media and there is elimination of contaminating agents (Johri and Mathew, 1987).

Modified techniques as use of alginate beads for entrapment of root fragments in alginate beads increases the quantity, quality and infectivity of the inoculum (Strullu and Plenchette, 1990). A system was recently adapted to allow the non-invasive microscopic observation of the symbiosis development between *Medicago truncatula* and *Glomus intraradices*. This "Mini-Mycorrhizotron" allowed the natural growth of the *Medicago* seedlings and permitted the observation of the fungal development in the first stage of the symbiosis (Bonanomi *et al*, 2001). However like any other *in vitro* system this one has some limitations as the need for special and large equipments and there is difficulty in managing such system.

Aeroponic AM culture production

(For a coloured version of this figure, see plate section, page 302)

iii) *Root organ culture:* Root organ culturing for axenic propagation of *Glomus mosseae* in clover root organ cultures was first performed by Mosse and Hepper (1975). The technique includes use of a modified agar based medium (MS rooting medium)/ liquid medium for creation of increased amount of roots from callus tissue and these roots are infected by AM spores or by surface sterilized root bits obtained from mycorrhizal plant. The surface sterilization of the infected root bits or the AM spores is performed by using common sterilants as 2% ChloramineT and 200ppm streptomycin, 0.5% sodium hypochlorite or 0.1% $MgCl_2$. However the intraradical hyphal forms of the fungi as mycorrhizal root pieces and single isolated vesicles may constitute excellent sources of inoculum for the establishment of *in vitro* cultures and for the continuous culture of the species. This technique is useful for production of contaminant free AM inoculum in greater amount and production of effective propagules, which could be recycled and used for inoculating micropropagated plants.

There are several variants for this basic technique which include protocols for obtaining large amount of root tissue to be utilized for AM colonization after infection as use of *Agrobacterium rhizogenes* Ri T-DNA transformed roots of plants which increases the rate of root production and thus quantities of VAM fungal propagules. Several workers reported the successful use of the above technique for mass production of AM fungal inoculum using transformed roots of tomato (Simoneau *et al,* 1994) and RiT-DNA transformed carrot roots (Declerck *et al,* 1996).

Gadkar and Adholeya (2000) obtained hairy roots by transformation of the host by RiT-DNA of *Agrobacterium rhizogenes* growing on a defined synthetic medium, which was then inoculated with mycorrhizal fungus. Guleri *et al* (2001) performed *in vitro* inoculation of micropropagated plants of sugarcane (CoJvar.86) with *Glomus fasciculatum* and observed minimum of 70 to maximum of 90 per cent VAM colonization in roots after 30 days of co-cultivation. Mycorrhizal plants were found to show increase in root biomass by two times in comparison to non-VAM infected plants. Further higher survival rate, sugar content and cane number was also reported in mycorrhizal micropropagated sugarcane plants. Sandhu *et al* (2001) observed that mycorrhizal infection was found to be higher when VAM fungi was exposed to stress conditions and there was increase in the number of asbuscules and vesicle after the stress. Mathur and Vyas (1999) established efficient symbiosis of *in vitro* multiplied VAM strain (*Glomus constructum*) with *in vitro* raised plantlets of *Ziziphus mauritiana* and observed significantly improved biomass production, nutrient uptake and acclimatization of the *in vitro* produced plantlets of *Z. mauritiana* into the field.

Aberrant in vitro derived plantlets

- poor photosynthetic activity (due to poor granal organization and etiolation effect due to ethylene gas)
- impaired stomatal behaviour (due to high Na^{+}/Ca^{2+} ions and incorrect micellation of the guard cells leading to decreased stomatal movements (enhanced dehydration)
- root system delicate and lack of root hair development (decreased nutrient and water absorption)
- poor vascular connection between root and shoot tissues (decreased water transportation)
- microbiological vacuum for the roots (enhanced chances of debacle in root system due to attack by phytopathogens)

iv) *Continuous culture:* Continuous cultures of AM fungal species could be obtained by regular and timely transfer of active AM fungal infected root tissue exhibiting mycorrhizal association to fresh medium containing or lacking spores and this root tissue proliferates again all over the medium. Since AM fungi have higher affinity for young root tissue thus if the infected root system becomes too old then fresh medium as well as fresh/actively growing root tissue is provided for maintaining the continuous culture.

For the maintenance of continuous cultures of AM fungi that produce less number of spores or vesicles as *Gigaspora* and *Scutellospora* sp., direct subculturing remains more difficult to achieve. The subcultured sample, a square (10–15 cm^2) of nondisturbed mycorrhizal roots and external mycelium, is transferred to a petri dish containing fresh medium.

Artificial in vitro system

- ❐ Ease in investigating the morphological changes microscopically
- ❐ Allows estimation and comparison of physiological activities among mycorrhizal and non-mycorrhizal plants.
- ❐ Provides information on signaling events during AM symbiosis

Dual/multicompartment systems

- ❐ Bicompartmental petri dish having rich medium in proximal and mineral only media in distal one (St-Arnaud *et al,* 1996)
- ❐ stimulation of mycelial growth and a 10-fold increase in spore production in distal compartment due to pH change or altered nutrient availability

The main constraint with the AM fungal popularity and wider use is obtaining pure culture and then mass production of the same for use as biofertilizer. Several techniques have emerged for the same process with advancement in newer protocols to amend the drawbacks of the former method. The prerequisite for the inoculum is its higher infectivity and positive effect on inoculation of any crop plant. Moreover the sufficient quantity of AM propagules is also required to initiate/maintain an ensured contact with the radicle after its emergence. The AM inoculations could be performed either by placing AM propagules an inch below seed or by spore application on seeds. Preferentially AM fungal inoculations could increase the yield and financial benefits by wide spread use in vegetables and transplantable crops.

Piriformospora indica

- new symbiotically endophytic fungi (Singh *et al,* 2000)
- belongs to class Basidiomycota (Hymenomycetes) based on
  - 18S rDNA analysis and
  - ultra structure of septal pore
- could be grown axenically on defined Kaefer's/potato dextrose agar medium (Varma *et al,* 1999)
- forms characteristic structures in cortical cells

- intracellular branched/dense coiled hyphal structures
- vesicle and spore-like structures (large pear shaped chlamydospores(8-25 nuclei) in axenic cultures)
- this plant interacting fungus exhibits diverse host range including
- bryophytes (liverwort rhizoids)
- terrestrial orchids particularly variuos *Dactylorhiza* species including *D. purpurella, D. incarnata* L, *D. majalis, D. fuchsii*
- angiosperms including both herbaceous plants (both monocots and dicots) and large woody trees
- plant growth-promoting root endophyte
- tremendously improves growth and biomass production of different plants
- promising candidate for biohardening of micropropagated plants
- functions similar to AM fungi i.e. acquisition of phosphates away from the depletion zone and absorption of minerals (micronutrients as zinc, iron, molybdenum etc
- induces resistance to fungal diseases (Walker *et al*, 2005)
- resistance to *Fusarium culmorum* and
- systemic resistance for *B.graminis* f.sp. *hordei* in barley
- help in increasing abiotic stress resistance/tolerance

Conclusions

AM fungal associations cherish elaborate benefits to terrestrial plants belonging to almost all known classes of land plants. This association requires prior signal transduction processes and result in inter and further intracellular invasion and inhabitation of fungal hyphae and formation of a variety of internal fungal structures for completion of process of P mobilization away from the depletion zone of the plant roots. Apart from providing host the required P AM fungal extramatricular hyphae also provide micronutrients as copper, zinc, molybdenum etc from the soil which are required as cofactors in variety of enzyme systems. AM fungal association also result in avoidance of abiotic stresses (drought, salt) as well as provide tolerance under extreme physiological conditions. Mycorrhizal plants are generally regarded more resistant to several plant pathogenic attacks than their non-mycorrhizal counterparts. AM fungi thus form the foremost armory in the brigade of biofertilizers and owing to the benefits these microbes confer to plants, their wider use and mass production is prerequisite in this era of organic/sustainable food production.

References

Asai T (1944) Uberdie Mycorrhizenbilidung der leguminosen pflanzen. *Japanese J Bot*, 13:463-485.

Augé RM (2004) Arbuscular mycorrhizae and soil/plant water relations. *Can J Soil Sci*, 84: 373–381.

Bakshi BK (1974) Mycorrhiza and its role in forestry. P.L. 480 Project Report. Forest Research Institute and Colleges, Dehra Dun, pp. 89.

Barker SJ, Tagu D and Delp G (1998) Regulation of root and fungal morphogenesis in mycorrhizal symbiosis. *Plant Physiol*, 116: 1201-1207.

Beck-Nielsen D and Madsen TV (2001) Occurrence of vesicular-arbuscular mycorrhiza in aquatic macrophytes from lakes and streams. *Aquatic Botany*, 71:141–148.

Bethlenfalvay GJ and Schuepp H (1994) Arbuscular mycorrhizas and agrosystem stability. *In: Impact of arbuscular mycorrhizas on sustainable agriculture and natural ecosystems*, Gianinazzi S and Schuepp H (eds.) Basel, Switzerland, Birkhauser Verlag, pp. 117–131.

Bonanomi A, Oetiker JH, Guggenheim R, Boller T, Wiemken A and Vögeli-Lange R (2001) Arbuscular mycorrhiza in mini-mycorrhizotrons: first contact of *Medicago truncatula* roots with *Glomus intraradices* induces chalcone synthase. *New Phytol*, 150:573-582.

Brewster JL and Tinker PBH (1972) Nutrient flow rates into roots. *Soils Fertilizers*, 35:355-359.

Butler EJ (1939) The occurrence and systematic position of vesicular - arbuscular type of mycorrhizal fungi. *Transactions British Mycol Soc*, 22:274-301.

Cairney JWG and Meharg AA (1999) Influences of anthropogenic pollution on mycorrhizal fungal communities. *Environ Pollut,* 106:169–182.

Chang MM, Culley DE, Hadwiger LA (1993) Nucleotide sequence of a pea (*Pisum sativum* L.) â-1,3-glucanase gene. *Plant Physiol,* 101:1121-1122.

Dalpe Y and Monreal M (2004) Arbuscular mycorrhiza inoculum to support sustainable cropping systems. Cited Online. Crop Management doi:10.1094/CM-2004-0301-09-RV.

De Bary A (1887) Comparative morphology and biology of the fungi, mycetozoa, and bacteria (Oxford, Clarendon Press).

Declerck S, Strullu DG and Plenchettc (1996) *In vitro* mass production of the arbuscular mycorrhizal fungus *Glomus versiforme,* associate with Ri T-DNA transformed carrot roots. *Myco Res,* 100(10): 1237-42.

Dietz KJ, Baier M and Kramer U (1999) Free radicals and reactive oxygen species as mediators of heavy metal toxicity in plants. *In:* Heavy metal stress in plants: from molecules to ecosystems Prasad MNN, Hagemeyer J (eds.), Berlin, Springer-Verlag, pp. 73-97.

Douds DD and Chaney WR (1982) Correlation of fungal morphology and development of host growth in green ash mycorrhiza. *New Phytol,* 92:519-526.

Douds Jr. DD, Nagahashi G, Pfeffer PE, Kayser WM and Reider C (2005) On-farm production and utilization of arbuscular mycorrhizal fungus inoculum. *Can J Plant Sci,* 85:15-21.

Duponnois R, Colombet A, Hien V and Thioulouse J (2005) The mycorrhizal fungus *Glomus intraradices* and rock phosphate amendment influence plant growth and microbial activity in the rhizosphere of *Acacia holosericea. Soil Biol Biochem,* 37:1460–1468.

El-Kherbawy M, Angle JS, Heggo A and Chaney RL (1989) Soil pH, rhizobia, and vascular-arbuscular mycorrhizae inoculation effects on growth and heavy metal uptake of alfalfa (*Medicago sativa* L.). *Biol Fert Soils,* 8:61-65.

Ferrol N, Barea JM and Azcon-Aguilar C (2000) The plasma membrane H^+-ATPase gene family in the arbuscular mycorrhizal fungus *Glomus mosseae. Curr Genet,* 37:112-118.

Fortin JA, Becard G, Declaerck S, Dalpe Y, St-Arnaud M, Coughlan AP and Piche Y (2002) Arbuscular mycorrhiza on root-organ cultures. *Can J Bot,* 80:1-20. Online citation DOI: 10.1139/B01-139

Gadkar V and Adholeya A (2000) Extraradical sporulation of AM *Gigaspora margarita* in long-term axenic cultivation in Ri T-DNA carrot root. *Mycological Res,* 104(6):716-21.

Gadkar V, Adholeya A and Satyanarayana T (1997) Randomly amplified polymorphic DNA using the M13 core sequence of the vesiculararbuscular mycorrhizal fungi *Gigaspora margarita* and *Gigaspora gigantea. Cana J Microbiol,* 43:795-798.

Gaur A and Adholeya A (2004) Prospects of arbuscular mycorrhizal fungi in phytoremediation of heavy metal contaminated soils. *Curr Sci,* 86(4):528-534.

Gerdemann JW and Bakshi BK (1976) Endogonaceae of India: two new species. *Trans Br Mycol Soc,* 66:340–343.

Giovannetti M, Avio L, Sbrana C and Citernesi AS (1993) Factors affecting appressorium development in the vesicular-arbuscular mycorrhizal fungus *Glomus mosseae* (Nicol. & Gerd.) Gerd. & Trappe. *New Phytologist,* 123:115–122.

Giovannetti M, Azzolini D and Citernesi AS (1999) Anastomosis formation and nuclear and protoplasmic exchange in Arbuscular Mycorrhizal fungi. *Appl Environ Microbiol,* 65:5571-5575.

Giovannetti M, Sbrana C, Avio L, Citernesi AS and Logi C (1993) Differential hyphal morphogenesis in arbuscular mycorrhizal fungi during pre-infection stages. *New Phytologist,* 125:587-593.

Graham JH and Abbott LK (2000) Wheat responses to aggressive and non-aggressive arbuscular mycorrhizal fungi. *Plant Soil,* 220: 207–218.

Guleri R, Gupta RP, Gosal SK, Pandher MS and Gosal SS (2001) *In vitro* Vesicular arbuscular mycorrhization of micropropagated sugarcane plants. Conference on Biotechnology -the science and the Business 28-30 Sept. .AIBA, Indian Institute of Technology , New Delhi, India

Gupta RP, Kalia A, Guleri R Pandher MS, Hunjan RK Gosal SK & Gosal SS (2007) Vesicular arbuscular mycotrriza. In: Organic recycling

& bioinoculans for sustainable crop production, eds LL Sonani and SC Bhandari, pp. 112-122.

Hall JL (2002) Cellular mechanisms for heavy metal detoxification and tolerance. *J Expt Bot,* 53(366):1-11.

Harrison MJ (1997) The arbuscular mycorrhizal symbiosis: an underground association. *Trends Plant Sci,* 2:54-60.

Harrison MJ (1999) Molecular and Cellular Aspects of the Arbuscular Mycorrhizal Symbiosis. *Annu Rev Plant Physiol Plant Mol Biol,* 50:361-389.

Harrison MJ and Van Buuren ML (1996) A phosphate transporter form mycorrhizal fungus *Glomus versiforme. Nature,* 378:626-629.

Helgason T Fitter AH and Young JPW (1999) Molecular diversity of arbuscular mycorrhizal fungi colonising *Hyacinthoides non-scripta* (bluebell) in a seminatural woodland. *Mol Ecol,* 8:659-666.

Johri BN and Mathew J (1987) Focal theme (Botany), *ISCA Symposium,* pp. 293-304.

Jones MD, Durall DM and Tinker PB (1998) A comparison of arbuscular and ectomycorrhizal *Eucalyptus coccifera*: growth response, phosphorus uptake efficiency and external hyphal production. *New Phytologist,* 140:125-134.

Kaldorf M, Zimmer W and Bothe H (1994) Genetic evidence for the occurrence of assimilarity nitrate reductase in arbuscular mycorrhizal and other fungi. *Mycorrhiza,* 5:23-28.

Kaya C, Higgs D, Kirnak H and Tas I (2003) Mycorrhizal colonization improves fruit yield and water use efficiency in watermelon (*Citrullus lanatus* Thunb) grown under well-watered and water-stressed conditions. *Plant Soil,* 253(2):287-292.

Khade SW and Rodrigues BF (2003) Occurrence of arbuscular mycorrhizal fungi in tree species from Western Ghats of Goa, India. *J Trop Forest Sci,* 15:320–331.

Killhan K and Firestone MK (1983) Vesicular arbuscular mycorrhizal mediation of grass response to acidic and heavy metal depositions. *Plant Soil,* 72:39-48.

Koide RT and Mosse B (2004) A history of research on arbuscular mycorrhiza. *Mycorrhiza,* 14:145-163.

Lanfranco L, Bianciotto V, Lumini E, Souza M, Morton JB and Bonfante P (2001) A combined morphological and molecular approach to characterize isolates of arbuscular mycorrhizal fungi in *Gigaspora* (Glomales). *New Phytol,* 152:169-179.

Lanfranco L, Delpero M and Bonfante P (1999) Intrasporal variability of ribosomal sequences in the endomycorrhizal fungus Gigaspora margarita. *Mol Ecol,* 8:37-45.

Lanfranco L, Wyss P, Marzachi C and Bonfante P (1995) Generation of RADP-PCR primers for the identification of isolates of *Glomus mosseae,* an arbuscular mycorrhizal fungus. *Mol Ecol,* 4:61-68.

Leyval C, Turnau K and Haselwandter K (1997) Effect of heavy metal pollution on mycorrhizal colonization and function: physiological, ecological and applied aspects. *Mycorrhiza,* 7:139–153.

Li XL and Christie P (2001) Changes in soil solution Zn and pH and uptake of Zn by arbuscular mycorrhizal red clover in Zn-contaminated soil. *Chemosphere,* 42:201–207.

Malcova R, Vosatka M and Gryndler M (2003) Effects of inoculation with *Glomus intraradices* on lead uptake by *Zea mays* L. and *Agrostiscapillaris* L. *Appl Soil Ecol,* 23(1): 255 - 267.

Manoharachary C, Sridhar K, Singh R, Adholeya A, Suryanarayanan TS, Rawat S and Johri BN (2005) Fungal biodiversity: Distribution, conservation and prospecting of fungi from India. *Curr Sci,* 89 (1): 58-71.

Mathur N and Vyas A (1999) Improved biomass production, nutrient uptake and establishment of *in vitro* raised *Ziziphus mauritiana* by VA Mycorrhiza. *J Plant Physiol,* 155:129 -32.

Meharg A, Bailey J, Breadmore K and McNair MR (1994) Biomass allocation, phosphorusnutrient and vesicular arbuscular mycorrhizal infections in clones of yorkshire fog, *Holcus lanatus* L that differ in their PO^{-}_{4} uptake kinetics and tolerance to arsenate. *Plant Soil,* 160:11-20.

Meharg AA (1994) Integrated tolerance mechanisms: constitutive and adaptive plant responses to elevated metal concentrations in the environment. *Plant Cell Environ,* 17: 989-993.

Merryweather J and Fitter A (1996) Phosphors nutrients of an obligately mycorrhizal plant treated with fungicide benomyl in field. *New Phytologist,* 132:307-311.

Mertz SM, Heithaus JJ and Bush RL (1979) Mass production of axenic spores of the endomycorrhizal fungus *Gigaspora margarita*. *Trans Br Mycol Soc*, 72:167-169.

Morandi D (1996) Occurrence of phytoalexins and phenolic compounds in endomycorrhizal interactions, and their potential role in biological control. *Plant Soil*, 185:241-251.

Morton JB and Benny GL (1990) Revised classification of arbuscular mycorrhizal fungi (Zygomycetes): a new order, Glomales, two new suborders, Glomineae and Giagasporineae, and two new families, Acaulosporaceae and Gigasporaceae, with an emendation of Glomaceae. *Mycotaxon*, 37:471-491.

Morton JB, Bentivenga SP and Wheeler WW (1993) Germplasm in the International Collection of Arbuscular and Vesicular-arbuscular Mycorrhizal Fungi (INVAM) and procedures for culture development, documentation and storage. *Mycotaxon*, 48:491-528.

Mosse B (1956) Fructifications of an Endogone species causing endotrophic mycorrhiza in fruit plants. *Ann Bot*, 20:349–362.

Mosse B (1990) Practical problems of field inoculation with VAM: Inoculum production. *Mycorrhiza News*, 2:2-4.

Mosse B and Hepper CM (1975) Vesicular arbuscular mycorrhizal infection in root organ culture *Physiol Plant Pathol*, 5:215 -23.

Mosse B and Thompson HP (1979) Improvements in or relating to the production of mycorrhizal fungi. *UK-Patent 2 043 688 A*.

Mosse B and Hayman DS (1971) Plant growth response to vesicular arbuscular mycorrhiza. II in unsterilized field soil. *New Phytologist*, 70:29-34.

Mosse B and Thompson JP (1983) Vesicular arbuscular mycorrhizal inoculum production 1. Exploratory experiments with been in nutrient flow culture. *Can J Bot*, 62:1523-30

Norman JR and Hooker JE (2000) Sporulation of *Phytophthora fragariae* shows greater stimulation by exudates of non-mycorrhizal than by mycorrhizal strawberry roots. *Mycol Res*, 104:1069-1073.

Peterson RL and Farquhar ML (1996) Root hairs: specialized tubular cells extending root surfaces. *The Botanical Revs*, 62:1-40.

Peyronel B (1923) Fructification de l'endophyte arbuscules et vesicules des mycorhizes endotrophes. *Bull Soc Mycol Fr,* 39:119–126

Pfeffer PE, Douds Jr DD, Becard G and Hill YS (1999) Carbon uptake and the metabolism and transport of lipids in an arbuscular mycorrhiza. *Plant Physiol,* 120:587-598.

Pirozynski KA and Dalpe Y (1989) Geological history of the *Glomeaceae,* with particular reference to mycorrhizal symbiosis. *Symbiosis, 7:* 1-36.

Quilambo OA (2000) Functioning of peanut (*Arachis hypogaea* L.) under nutrient deficiency and drought stress in relation to symbiotic associations. PhD thesis. University of Groningen, the Netherlands. Van Denderen BV, Groningen.ISBN 903671284X.

Rabie GH (2005a) Contribution of arbuscualr mycorrhizal fungi to red kidney and wheat plant tolerance grown in heavy metal polluted site. *Afr J Biotechnol,* 4(4):332-345.

Rabie GH (2005b) Role of arbuscular mycorrhizal fungi in phytoremediation of soil rhizosphere spiked with poly aromatic hydrocarbons. *Mycobiol,* 33(1):41-50.

Redecker D, Morton JB and Bruns TD (2000) Ancestral lineages of arbuscular mycorrhizal fungi (Glomales). *Mol Phylogen Evol,* 14:276-284.

Redecker D, Thierfelder H, Walker C and Werner D (1997) Restriction Analysis of PCR-Amplified Internal Transcribed Spacers of Ribosomal DNA as a tool for species identification in different genera of the order Glomales. *Appl Environ Microbiol,* 63: 1756-1761.

Rosendahl S and Taylor JW (1997) Development of multiple genetic markers for studies of genetic variation in arbuscular mycorrhizal fungi using AFLPTM. *Mol Ecol,* 6:821-829.

Ruiz-Lozano JM (2003) Arbuscular mycorrhizal symbiosis and alleviation of osmotic stress: New perspectives for molecular studies. *Mycorrhiza,* 13:309-317.

Ruiz-Lozano JM, Collados C, Porcel R, Azcon R and Barea JM (2002) Identification of a cDNA from the arbuscular mycorrhizal fungus *Glomus intraradices* that is expressed during mycorrhizal symbiosis and up-regulated by N fertilization. *MPMI,* 15(4):360-367.

Sandhu A, Gosal SK,Gupta RP and Gosal SS (2001) Effect of abiotic stresses on *in vitro* VAM multiplication. *Ibid* (Abste).

Schreiner RP and Bethlenfalvay GJ (1995) Mycorrhizal interactions in sustainable agriculture. *Crit Rev Biotechnol,* 15:271–285.

Selosse MA and Le Tacon F (1998) The land flora:a phototroph-fungus partnership? *Tree,* 13(1):15-20.

Shetty KG, Hetrick BA and Schwab AP (1995) Effects of mycorrhizae and fertilizer amendments on zinc tolerance of plants. *Environ Pollut,* 88:307-314.

Simon L, Levesque RC and Lalonde M (1993) Identification of endomycorrhizal fungi colonizing roots by fluorescent single strand conformation polymorphism-polymerase chain reaction. *Appl Environ Microbiol,* 59:4211–4215.

Simoneau P, Louis NL, Plenchette C and Strulle DG (1994) Accumulation of new polypeptides in Ri T-DNA transformed roots of tomato during the development of vesicular arbuscular mycorrhiza. *Appl Environ Micribiol,* 60(6):1810-13.

Singh A, Sharma J, Rexer KH and Varma A (2000) Plant productivity determinants beyond minerals, water and light: *Piriformospora indica*–A revolutionary plant growth promoting fungus. *Curr Sci,* 79(11):1548-1554.

Singh R and Adholeya A (2002) Biodiversity of arbuscular mycorrhizal fungi (AMF) in different agroclimatic regions of India. IMC 7-Conference on Mycological Advances, 11–17 August 2002, Norway.

Smith SE and Bowen GE (1979) Soil temperature, mycorrhizal infection and nodulation of *M. truiculata* and *Trifolium subteeraneum. Soil Biol Biochem,* 11:469-73.

Smith SE and Gianinazzi-Pearson V (1990) Phosphate uptake and arbuscular activity in mycorrhizal *Allium cepa* L. Effects of photon irradiance and phosphate nutrition. *Aust J Plant Physiol,* 17: 177–188

Smith SE and Read DJ (1997) The symbionts forming VA mycorrhiza. *In: Mycorrhizal Symbiosis* (2nd ed.), Acad. Press London, UK, pp.11-33.

St-Arnaud M, Hamel C, Vimard B, Caron M and Fortin JA (1995) Altered growth of *Fusarium oxysporum* f.sp. *chrysanthemi* in an *in vitro* dual culture system with the vesicular arbuscular mycorrhizal fungus *Glomus intraradices* growing on *Daucus carota* transformed roots. *Mycorrhiza,* 5:431-438.

Strullu DG and Plenchette C (1990) Encapsulation of intraradical form of *Glomus* in alginate and use of capsules as inoculum. *Cr Acad Sci III Vie,* 310:447-452.

Strullu DG, Diop T and Plenchlte C (1997) Constitution of in vitro collection: A proposed life cycle of *Glomus.* Comptes Rendus Del. Academic Des Sciences Series 111-Sciences De La Vie *Life Sciences,* 20(1):41-47.

Taylor TN, Hass H and Kerp H (1997) A cyanolichen from the Lower Devonian Rhynie chert. *Amer J Bot,* 84:992-1004.

van Rhijn P, Fang Y, Galili S, Shaul O, Atzmon N, Wininger S, Eshed Y, Lum M, Li Y, To V *et al* (1997) Expression of early nodulin genes in alfalfa mycorrhizae indicates that signal transduction pathways used in forming arbuscular mycorrhizae and *Rhizobium*-induced nodules may be conserved. *Proc Natl Acad Sci USA,* 94:5467-5472.

van Tuinen D, Jacquot E, Zhao B, Gollotte A and Gianinazzi-Pearson V (1998) Characterization of root colonization profiles by a microcosm community of arbuscular mycorrhizal fungi using 25S rDNA-targeted nested PCR. *Mol Ecol,* 7:879-887.

Varma A, Verma S, Sahay N, Butehorn B and Franken P (1999) *Piriformospora indica,* a cultivable plant-growth-promoting root endophyte. *Appl Environ Microbiol,* 65(6):2741-2744.

Walker F, Achatz B, Baltruschat H, Fodor J, Becker K, Fischer M, Heier T, Huckelhoven R, Neumann C, von Wettstein D, Franken P and Kogel KH (2005) The endophytic fungus *Piriformospora indica* reprograms barley to salt-tolerance, disease resistance and higher yield. *PNAS,* 102(38):13386-1191.

Weissenhorn I, Leyval C and Berthelin J (1995) Bioavailability of heavy metals and abundance of arbuscular mycorrhiza in a soil polluted by atmospheric deposition from a smelter. *Biol Fertil Soil,* 19:22-28.

Weissenhorn I., Leyval C, Belgy G and Berthelin J (1995) Arbuscular mycorrhizal contribution to heavy metal uptake by maize (*Zea mays* L.) in pot culture with contaminated soil. *Mycorrhiza,* 5: 245–251.

Yu X, Cheng J and Wong M H (2004) Earthworm-mycorrhiza interaction on Cd uptake and growth of ryegrass. *Soil Biol Biochem,* 37:1-7.

Mind Bogglers

1. What are AM fungal associations? How these associations were helpful in smoother movement and spread of plants from aquatic to terrestrial life?
2. Why AM fungi are termed obligate biotrophs? What are the major groups and genera of these fungi?
3. The molecular systematic approaches have resulted in reallocation of several genera and species in newer taxons. Justify
4. What are the events in the life cycle of AM fungi that are signal regulated? Why initially the hyphae remain apoplastic and finally invade the host cytoplasm?
5. How AM fungal extramatricular hyphae scavenge available P from farther regions in soil and enrich P nutrition of inoculated plants?
6. Describe the other benefits of AM fungal associations apart from mineral nutrition.
7. What are problems of AM fungal culture and mass production? What according to you would be the best technique for obtaining large-scale production of AM fungal inoculum?
8. How AM fungi protect plant from pathogenic fungi but not viruses?
9. What are different soil less media culture methods for multiplication of AM fungi?
10. Why AM fungal association is not observed in members of brassicaceae family?

11. What is the role of VAS elements in life cycle and events of infection in AM fungi?
12. What are the probable alternative candidate (s) to AM fungi for enrichment and supply of macro and micronutrients to plants? How these facultative endophytes help in developing stress tolerance and disease resistance in biotized or inoculated plants?

7

Plant Growth Promoting Rhizobacteria and Biocontrol Agents

- Introduction 184
- Mechanisms of Plant Growth Stimulation 186
- Mechanisms of Biocontrol Action (Bioprotectant) 189
- Future Perspectives 195
- References 198
- Mind Bogglers 200

INTRODUCTION

Lithosphere, including the earth's crust, is helpful in providing the base or medium for sustenance of most life processes and forms, interwoven in fine network interdependencies within themselves and the environment surrounding them i.e. ecosystems. Terrestrial ecosystem contains an important dynamic living matrix termed soil, which is critical for agricultural production and food security. Soil forms the medium for growth and maintenance of a variety of plant species while the plant roots in the soil represent a four dimensional region in space and time of profuse activity relative to the bulk soil, revolving around pH, nutrient, redox potential, and exudate gradients changing as distance from the root increases (Marschner, 1995). The soil region adhering the plant roots is being strongly influenced by the presence of these roots and is characterized by high rates of microbial population and activity and referred as rhizosphere (the soil adhering to plant roots).

The plant roots constantly alter the rhizospheric soil environment by secreting root border cells and root exudates thereby creating a rich source of diverse, abundant, and reliable substrates in extremely heterogeneous adjoining soil environment in terms of nutrients and water supply through time and space. These exudates consist of a wide array of compounds and simple substrates from low molecular weight compounds such as sugars, phenolics, amino acids, organic acids, and other secondary metabolites (more than 100,000 compositionally different compounds) to higher molecular weight compounds such as proteins and mucilage.

Soil is considered a storehouse of microbial activity, though the space occupied by living microorganisms is estimated to be less than 5% of the total space. The exact nature of compounds in root exudates has a pivotal role in activation of microbial genes responsible for recognition and initiation of symbiotic association or these may act as an antimicrobial plant defense, may activate or disrupt key microbial genes responsible for biofilm formation, or simply may act as an easy source of moisture, nutrients, and energy. The rhizosphere effect on soil microbes is highly reciprocal and dynamic and could be attributed to mutual influx and efflux activities of plant and rhizomicrobial population primarily including root exudates (complex organic and chemical phenolics) and mineral diffusion by the two counterparts respectively. It could be commented that the plants provide the major source of carbon for maintenance of microbial community in the rhizosphere and the quantity and nature of root exudates determine the composition of rhizosphere microflora. Moreover, the plant itself depends on the ability of microbial community to make required nutrients available including nitrogen, phosphorus and iron.

The rhizosphere microbial diversity and population may exhibit close association with plants as these versatile agents have developed efficient portal systems to circumvent or subvert the defense protocols or processes of plants which may be detrimental or beneficial to plants. These diverse rhizomicroflora members can exhibit both synergistic, additive, competitive or antagnostic interactions that elaborately affect plant growth. Competition for nutrients, production of antibiotics or bacteriocins, production of siderophores and secretion of lytic enzymes or secondary metabolites are mechanisms which permit some microbial population to be sustained, eliminated or favoured. Such interactions can determine the development or suppression of certain plant diseases.

Some soil bacteria preferentially associate with the roots of crop plants and can exert beneficial effects on their hosts. They are collectively referred to as PGPR (Plant Growth Promoting Rhizobacteria). Thus these bacteria form that subset of soil and rhizosphere bacteria that colonize roots in a competitive environment as soil rhizosphere and rhizoplane regions. PGPR strains are often found among the fluorescent pseudomonads, *Acetobacter, Actinoplanes, Agrobacterium,*

Alcaligenes, Arthrobacter, Azospirillum, Azotobacter, Bacillus, Cellulomonas, Enterobacter, Erwinia, Flavobacterium, Pasteuria, Serratia and *Xanthomonas*. The other group of beneficial bacteria especially rhizobia and bradyrhizobia form a symbiotic relationship with legume plants and have been used extensively worldwide as legume inoculants for increasing crop yields.

Factors responsible for stimulation of plant growth: Rhizobacteria are used extensively for inoculation of agricultural crops, which are able to colonize plant roots and stimulate plant growth as well as crop yield. There are several mechanisms by which these rhizobacteria enhance plant growth (being multitudinous) and include

(i) the ability to produce or change the concentration of the plant hormones like indoleacetic acid (IAA), gibberellic acid, cytokinins, and ethylene;

(ii) asymbiotic N_2 fixation

(iii) antagonism against phytopathogenic microorganisms (e.g., *Fusarium* spp by production of siderophores, ß-1,3-glucanase, chitinases, antibiotics, and cyanide)

(iv) solubilization of mineral phosphates and other nutrients.

Mechanism of Plant Growth Stimulation

(i) *Production of plant growth promoting substances (Biostimulant):* Many root colonizing bacteria including *Pseudomonas* spp are known to produce vitamins and growth hormones (auxins, cytokinins etc), which often lead to morphological changes in the plants. Thus, these bacterial strains regulate plant growth by mimicking synthesis of plant hormones including indole-3-acetic acid, cytokinins, or gibberellins, while others increase mineral and nitrogen availability in the soil. Plants differ in the levels and ratio of the hormones required to maintain normal growth and development therefore, it might be expected that different stages of development respond differently to invasion by hormone producing bacteria. It is generally believed that hormone production by rhizobia plays a role in development and maintenance of nodules. *Bradyrhizobium elkanii* strains have been observed to synthesize and secrete IAA in culture. IAA production by *R. meliloti* is stimulated by addition of flavonoids. *Azospirillum brasilense* was found to produce extremely high amounts of IAA. However, the amounts of IAA produced depended on species and strain as well as on the conditions of cultivation. Inoculation with *Azospirillum* increases the density and length of root hairs, as well as the appearance and elongation rates of lateral roots, thus increasing the root surface area.

Pseudomonas savastanoi has been reported to produce large amounts of IAA when grown in presence of tryptophan. This bacterium causes tumor like outgrowths in olives. The capacity to induce tumors is correlated with capacity to produce auxins which indicates that auxin is major determinant in virulence and also suggest that microbially released plant growth promoting substances in the rhizosphere affect plant growth.

Glick *et al* (1998) have reported another mechanism of plant root growth stimulation that is plant production of 1-aminocyclopropane-1-carboxylate (ACC) in response to bacterial production of IAA by *Pseudomonas putida* GR12-2 that may stimulate plant cell proliferation or elongation. ACC is a precursor of gaseous phytohormone ethylene that inhibits root elongation of seedlings. The plant-produced ACC is taken up by the PGPR strain and is hydrolyzed by ACC deaminase, thereby lowering the pool of ACC and reducing the level of ethylene. The net biological effects of this system are increased root elongation of the plant and a nitrogen source for the PGPR.

(ii) *Increased availability and uptake of nutrients (Biofertilizer):* A number of PGPR including pseudomonads, *Enterobacter, Acetobacter, Beijerinckia* and *Klebsiella* can fix atmospheric nitrogen and contribute a significant input of nitrogen to plants. Inoculation of cereal plants with nitrogen fixing bacteria has been found to cause increase in plant dry matter along with increased N percentage in plant tissue of sorghum, millet, maize, barley, rice and wheat indicating effects of inoculation on nitrogen fixation and N assimilation (Steenhoudt and Vanderleyden, 2000). PGPR are also found to stimulate plant growth by facilitating the uptake of minerals into the plant roots. It has been observed that PGPR inoculated plants have increased N, P, K content at different stages of plant development in sorghum, corn, wheat and millet. PGPR strains have been found to increase the root length and root biomass and the better developed root system may increase the mineral uptake in plants (Blomberg and Lugtenberg, 2001). Rodrýguez and Fraga (1999) observed increased P-solubilization by certain microbial types and inoculation of these microbes resulted in increased P-availability and uptake by inoculated plants.

Canola (*Brassica napus* L. cv) productivity decreases particularly in sulphur deficient soils (Scherer, 2001).

Microbial Sulphur Conversion

- slow process
- involves oxidation of elemental S to SO42-
- accelerated by inoculation of S-oxidizing bacteria
  Banerjee & Yesmin (2002)

Banerjee and Yesmin (2004) have reported significant increase in the performance and production of canola by use of sulphur oxidizing bacterial inoculant *Delftia acidovorans* (BioBoost) isolated from Canadian soil, which also possesses PGPR abilities. Seed analysis showed that BioBoost inoculant helped in canola S-uptake but did not change the seed quality traits like oil, protein, oleic acid, linolenic acid and glucosinolate content of canola seed.

(iii) *Siderophore production:* Iron is an essential trace element and is generally abundant in soil. Due to its very low solubility at neutral and alkaline pH values, iron is often unavailable to both plants and microorganisms. To cope with its insolubility, many organisms have evolved high affinity iron uptake of ferric siderophores. Siderophores are low molecular weight, high affinity Fe^{3+} chelators that transport iron into bacterial cells. In the rhizosphere siderophores produced by biocontrol agents are able to sequester the limited supply of iron thereby making it unavailable to pathogenic microbes and restricting their growth. Iron competition as a mechanism of biological control can be complicated as some siderophores can only be used by bacteria that produce them whereas others siderophores can be used by different bacteria. Iron competition is further complicated by the fact that certain siderophouse as pyoverdin and salicylate may act as elicitors for inducing systemic resistance (Whipps, 2001).

Various plant associated rhizosphere bacteria have been found to produce different types of siderophores. The siderophore producing PGPR such as *Azotobacter, Pseudomonas, Rhizobium* and *Azospirillum* improve the availability of iron nutrition of plants by making the unavailable form of iron to available form besides siderophore mediated competition for iron also confers competitive advantage to producer bacteria in interactions in the rhizosphere.

Siderophore produced by plant associated rhizosphere bacteria

Siderophore	Producing bacteria
a. Pyoverdin	*Pseudomonas spp, P. fluorescens, P. putida*
b. Catechols	
agrobactin	*Agrobacterium tumifaciens*
Enterobactin	Enterobactreiaceae family
Azotochelin	*Azotobacter vinelandii*
c. Other types	
Rhizobactin	*Rhizobium meliloti*
Citric acid	*Bradyrhizobium japonicum, Bradyrhizobium sp.*
Azotobactin	*Azotobacter vinelandii*

Mechanism of Biocontrol Action (Bioprotectant)

The bioprotecant action of PGPRs is due to production of a variety of compounds/secondary metabolites and other products that culminates to eradication/suppression of the plant pathogenic forms in vicinity of plant roots. There are five mechanisms of bioprotecant action reported so far and generally a single bioprotecting strain possesses one or more properties or follows more than one kind of mechanism to combat the plant pathogenic microbes. However there are several reports that suggest the biocontrol of nematodes and many other pathogens by inoculation with PGPRs. Biological control can occur through different modes of action including antibiosis, competition for iron by production of siderophores, competition for colonization sites and nutrients, induced systemic resistance, inactivation of pathogen germination factors, degradation of pathogen toxins and parasitism that may involves the production of cell-wall degrading enzymes (Whipps, 2001).

Li *et al* (2005) have reported the biocontrol effect of inoculation of four rhizobacterial *Bacillus subtilis/Brevibacillus brevis* strains (isolated from vegetable rhizosphere in china) in suppression of root-knot nematode and soil-borne fungal pathogens of mungbean caused by *Meloidogyne javanica and Rhizoctonia solani* SX-6, *Pythium aphanidermatum* ZJP-1 and *Fusarium oxysporum* f.sp. *cucumerinum* ZJF-2 in laboratory and greenhouse conditions in Belgium.

Antagonism: Antagonism or amensalism is a phenomenon of negatively affecting the survivility of other organisms by direct predation or parasitism or by production of certain antagnostic compounds. The general trends are the competition for nutrients and niche, parasitism and antibiosis through production of antibiotics, siderphores and other metabolites.

Production of antibiotics and bacteriocins: Plant-microbial ecological studies have been prerequisite for developing plant disease control strategies. One of the major mechanisms postulated for the biological control of plant root diseases is the production of antimicrobial compounds by the disease control agent. An antibiotic is defined as organic chemical substance or compound produced by a microbe organism (bacteria or molds) in low concentrations that is capable of inhibiting or killing (susceptible) microorganisms or chemical compounds used to kill or inhibit the growth of infectious organisms. The PGP microbes basically combat the root pathogenic and soil borne pathogens. Antibiosis involves production of antifungal metabolites including ammonia, butyrolactones, 2-4-diacetylphloroglucinol, HCN, kanosamine, Oligomycin A, Oomycin A, phenazine-1-carboxylic acid, pyoluterin, pyrrolnitrin, viscosinamide, xanthbaccin and zwittermycin A as well as several uncharacterized moieties (Whipps, 2001). These act as antibiotics killing non-resistant bacteria and fungi. Antimicrobial compounds may act on plant pathogenic fungi by inducing fungistasis or inhibition of spore germination, lysis of fungal mycelia or by exerting fungicidal effects.

Brodhagen *et al* (2004) have reported the production of polyketides (secondary metabolites) pyoluteorin and 2,4-diacetylphloroglucinol by *Pseudomonas fluorescens* Pf-5 which were positively autoregulated that is addition of pyoluteorin alone to liquid cultures of Pf-5 enhanced its production. Addition of both compounds mutually inhibited production of one another demonstrated at the transcriptional level by measuring activity from transcriptional fusions of an ice nucleation reporter gene (*inaZ*) to three separate pyoluteorin biosynthetic genes. A similar positive autoregulation was observed under natural rhizospheric conditions suggesting the role of polyketides as signal molecules influencing the spectrum of secondary metabolites produced by the bacterial cell.

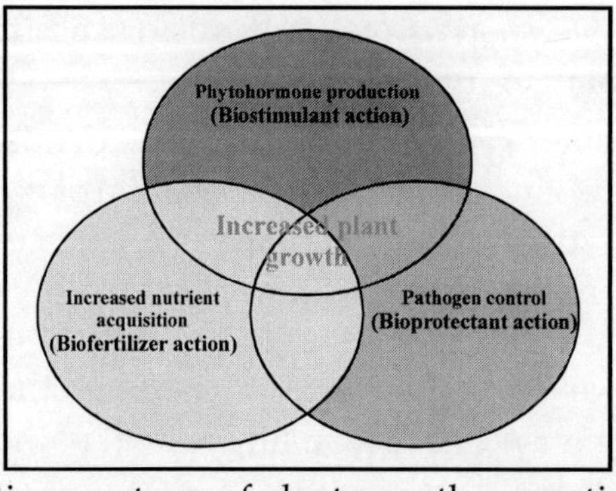

Mechanism/action spectrum of plant growth promoting rhizobacteria

Bacteriocins are proteinaceous antagnostic substances belonging to important class of antibiotics produced by bacteria that are lethal to other bacteria. These are usually peptides or proteins that selectively kill related bacteria (of the same species or genus) but do not affect other organisms. Bacteriocins are adsorbed to the receptor on the surface of bacterial cells, these eliminate electric charges of the inner membrane and stop biosynthesis of adenosine triphosphate thereby disrupting the biosynthesis of DNA, RNA, proteins and phospholipids and eventually growth of the bacterial cell and may lead to cell death. Bacteriocin-mediated antagonism is believed to occur in virtually any niche colonized by bacteria. Bacteriocin production is detectable in most strains of human opportunistic pathogen *Pseudomonas aeruginosa* and has been used to differentiate clinical isolates. Likewise, bacteriocin production is ubiquitous among plant-associated *Pseudomonas*. Bacteriocins producing strains have been identified among natural isolates of *R. trifolii, R. leguminosarum, R. japonicum* and cowpea rhizobia.

ii) *Production of secondary metabolites:* A number of alternate mechanisms have also been observed for increased plant growth and resistance to disease in plants inoculated with biocontrol bacteria. The production of potent extracellular lytic enzymes such as chitinases and laminarinase, capable of destroying fungal cell walls of *Fusarium* sp and other fungal pathogens, by some isolates of rhizospheric bacteria as *Pseudomonas stutzeri, Serratia marscens* etc (cultures or cell free extracts) is a possible mechanism of disease control.

El-Tarabily *et al* (1997) have reported this mechanism of action by PGP actinobacteria i.e. parasitism of pathogenic fungal spores by rhizospheric actinobacteria due to production of extracellular enzymes. A similar report from Valois *et al* (1996) established the production of α-1-3, α-1-4 and α-1-6 glucanases from a number of actinobacterial isolates that were able to suppress raspberry root rot causing fungal pathogen *Phytophora fragariae* var. *rubi*.

The other mechanism of biological disease control is the exclusion of minor pathogens or deleterious microorganisms from the rhizosphere by the beneficial bacteria. It may be due to production of toxic metabolites by rhizosphere bacteria, which inhibit the plant growth by damaging the roots. Potato rhizosphere pseudomands have been reported to produce hydrogen cyanide (HCN), which is harmful to root growth. Inhibition of fungal germination and growth by volatile compounds in the soil atmosphere has also been observed occasionally. These compounds probably result from the metabolic activities of soil bacteria. Ethylene, which inhibits the germination of fungal spores, has frequently been detected in the soil.

iii) *Nutrient competition and niche exclusion:* Competition of nutrients supplied by root and seed exudates occurs between beneficial rhizobacteria and pathogens on the root and is responsible to a small degree for the observed biocontrol by introduced bacteria. There is a correlation between the ability of various bacterial strains to compete for nutrients with germinating oospores of pathogenic fungi and their efficacy as biocontrol agents against pathogen cattack & cappearance of disease, which suggest that large populations of bacteria established on planting material and roots become a partial sink for nutrients in the rhizosphere, thus reducing the amount of carbon and nitrogen available to stimulate germination of spores of fungal pathogens or for subsequent colonization of the root. Fluorescent pseudomands are especially suited for rapid uptake or scavenging of nutrients, since they are nutritionally versatile and grow rapidly in the rhizosphere. Niche exclusion is potentially an important mechanism of antagonism of deleterious rhizobacteria by PGPR as the products secreted by PGPR exhibit exclusion of harmful rhizomicrobes via antagonistic interactions.

iv) *Production of phytoalexins or flavonoid-like compounds and induction of systemic resistance:* Phytoalexins are low molecular weight antimicrobial compounds, which are produced in the infected tissue by plants under stress conditions or in response to bacterial invasion. The phytoalexin production is also considered as an important mechanism of disease suppression in plant-pathogen relationships. At low concentration, flavonoid molecules secreted by plant roots can act as potential signal molecules causing selective chemotaxis followed by increased accumulation of certain rhizobacteria in rhizosphere. At higher concentration phytoalexins or flavonoids act as active defence molecules involved in induced systemic resistance.

Preinoculation of plant roots or seeds with avirulent or weakly aggressive strains of disease causing fungi results in induction of Induced systemic resistance (ISR) of plants against fungal diseases. The seed bacterization of common bean with *P.fluorescens* strain S97 induced disease resistance in bean against halo blight caused by *P.syringae* bv. *phaseolicola*. The seed inoculation of cucumber with selected PGPR strains induced resistance in cucumber leaves against anthracnose caused by *Colletotrichum orbiculare.* Strains of *Pseudomonas* spp. were also found to induce systemic resistance in cucumber against *Pythium aphanidermatum*. A root-colonizing *P. fluorescens* strain CHAO, when applied in soil, induced resistance in tobacco against tobacco necrosis virus (compant et al, 2005a).

v) *Suppression of growth of pathogenic microbes:* The suppression of growth of soil borne plant pathogens by use of microorganisms to reduce disease is termed as biocontrol. Microorganisms that can grow in the rhizosphere are ideal for use as biocontrol agents, since the rhizosphere provides the front line defense for roots against attack by pathogens. Biocontrol of phytopathogens appears to be a major mechanism of plant growth promotion by these bacteria. Suppression of phytopathogens results from production of secondary metabolites or by elicitation of the plant's own defense system. PGPR-based inocula must be able to compete with the indigenous microorganisms and efficiently colonize the rhizosphere of the plants to be protected. Rhizosphere microorganisms used to control diseases in various crops are shown below.

Bacterial biocontrol agents for specific plant pathogenic diseases

Biocontrol Organisms	Crop	Causal organism/disease
Pseudomonas fluorescens	Cotton Cotton Cotton Wheat Wheat	Damping off *Rhizoctonia solanii* *Pythium ultimum* Take-all *Gaeumannomyces graminis var. tritici*
Pseudomonas putida	Flex, radish, cucumber Beans Potato	Fusarium wilt *Fusarium solanii* *Erwinia carotovora, Xanthomonas compestris*
Bacillus subtilus	Corn	*Fusarium roseum*
Rhizobium and Bradyrhizobium	Soybean Mung bean Sunflower Okra	*Macrophomina phaseolina* *Rhizoctonia solanii* *Fusarium solanii*
Bacillus subtilis *Brevibacillus brevis*	Mungbean	*Meliodogyne javanica* (Nematode)

FUTURE PERSPECTIVES

Significant advances in elucidation of mechanisms involved in plant growth promotion have been made owing to advancements in molecular biology approaches, which mainly include the role of biological nitrogen fixation, production of phytohormones and biological control. The rhizosphere region of soil is swarmed by several microbial forms including the beneficial free-living PGPR that have pronounced role in plant growth enhancements by virtue of several mechanisms particularly useful for horticultural crops. Inspite of versatility of action spectrum for a visibly sound positive effect the use of PGPR has been most intensively researched in agriculture and horticulture though new fields of PGPR usage are emerging such as forest regeneration (Bisht *et al,* 2003) and phytoremediation of contaminated soils (Lucy *et al,* 2004). As the mechanisms of plant growth promotion by these bacteria are unravelled, the possibility of more efficient plant-bacteria pairings for novel and practical uses will follow.

Involvement of volatile organic components (VOCs) in bacterial signaling processes is the most recent hypothetical mechanism for plant-growth promotion by certain PGPR (Ping and Boland, 2004). This proposed mechanism underlines the fact that PGP bacterial VOC emissions trigger physiological plant growth promoting activities in the inoculated plant though the actual underlying molecular mechanisms responsible for growth promotion are not characterized. Ryu *et al* (2003) have performed microarray analysis of transcriptome of *Arabidopsis thaliana* for genes that were differentially expressed on exposure to bacterial VOC emissions and reported upregulation of a group of auxin-related genes and transcriptional regulation of selected gene was confirmed by reverse transcription-PCR.

Another field most likely to be very fruitful regarding use of PGPRs is biotization of micropropagated plants. Under soil less culture conditions stress may be caused by several abiotic factors as temperature, light, O_2/CO_2 concentrations, altered nutrient supply, drought, salinity as well as by biotic factors including pathogens. Since soil-less cultivation involves growing plants under aseptic conditions, it creates a microbiological vacuum for the plant roots and this calls for inoculation of beneficial microorganisms in different growing systems that may prevent either prophylactically or moderate the stressing situations (Woitke and Schitzler, 2005). The molecular biology of plant-microbial interaction and response has been providing answers to reason of biotic/abiotic stress omission/resistance in inoculated plants. A report by Timmusk and Wagner (1999) suggests a relative alteration in *Arabidopsis thaliana* plant gene expression confirmed by differential RNA display, induced by inoculation with isolates of PGP rhizobacteria *Paenibacillus polymyxa*. They also reported that these changes confer resistance to biotic (*Erwinia carotovora*) and abiotic stress (induction of drought) in the inoculated plants over control plants. The RNA analysis of the inoculated and uninoculated plants provided a strong co-relation of *ERD15* gene (drought stress responsive gene) expression with the inoculation of PGPR isolates suggesting that genes and/or gene classes associated with plant defenses against abiotic and biotic stress may be co-regulated.

There is an increasing interest in ascertaining the site of colonization of several PGPRs as the competitive environment of the soil rhizosphere and rhizoplane could result in low establishment and decreased benefits of the inoculated strains. Now researchers are more interested in isolating and characterizing strains that exhibit an endophytic relationship with the inoculated plant. Compant *et al* (2005b) have provided patterns of colonization of *Vitis vinifera* L. cv. Chardonnay plantlets by *Burkholderia* sp. PsJN under gnotobiotic conditions. By using *gfp/gusA* tagged strain (PsJN::*gfp*2x, PsJN::*gusA*11) they reported chronological occurrence of bacteria on root surfaces/cell wals/rhizodermal cell surface, at lateral root emergence sites and root tips, in root internal tissues and in the fifth internode and the tissues of the fifth leaf such as the substomatal chambers. *Burkholderia* sp. PsJN induces a local host defense reaction and systemically spreads to aerial parts through the transpiration stream.

Benhamou *et al* (1996) have given evidence at the ultrastructural level using an *in vitro* system towards the potential role of endophytic bacteria as inducers of plant disease resistance using *Bacillus pumilus* SE34 in root inducing T-DNA pea *(Pisum sativum* l.) infected with the pea root-rotting fungus *Fusarium oxysporum* f.sp. *pisi*. They reported restricted pathogen growth (at epidermis and outer cortex) in prebacterized plant roots due to induction of an array of host responses including strengthening of epidermal and cortical cell walls, enhanced callose deposition at the wall appositions, infiltration of wall appositions with phenolic compounds and deposition of newly formed barriers beyond the infection sites than in nonbacterized roots where the pathogen multiplied abundantly through much of the tissue including the vascular stele.

References

Banerjee MR and Yesmin L (2002) Sulfur oxidizing rhizobacteria: an innovative environment friendly soil biotechnological tool for better canola production. *In: Proceedings of AGROENVIRON 2002,* Cairo, Egypt, October 26-29, pp. 1-7.

Banerjee MR and Yesmin L (2004) BioBoost: a new sulfur-oxidizing bacterial inoculant for canola. *In: Proceedings of the 4th International Crop Science Congress* Brisbane, Australia, 26 Sep-1 Oct 2004, ISBN 1920842209.

Benhamou N, Kloepper JW, Quadt-Hallman A and Tzum S (1996) Induction of defense-related ultrastructural modifications in pea root tissues inoculated with endophytic bacteria. *Plant Physiol,* 112:91 9-929

Bisht D, Pandey A and Palni LMS (2003) Influence of microbial inoculations on *Cedrus deodara* in relation to survival, growth promotion and nutrient uptake of seedlings and general soil microflora. *J Sustainable Forestry,* 17(3):37-54.

Bloemberg GV and Lugtenberg BJJ (2001) Molecular basis of plant growth promotion and biocontrol by rhizobacteria. *Curr Opin Plant Biol,* 4:343-350.

Brodhagen M, Henkels MD and Loper JE (2004) Positive autoregulation and signaling properties of pyoluteorin, an antibiotic produced by the biological control organism *Pseudomonas fluorescens* Pf-5. *Appl Environ Microbiol,* 70(3):1758-1766.

Complant C, Duffy B, Nowak J, Clement C and Barka A (2005a) Use of plant growth-promoting bacteria for biocontrol of plant diseases: Principles, Mechanisms of action, and future prospects. Appl Environ Microbiol, 71(a): 4951-4959.

Compant S, Reiter B, Sessitsch A, Nowak J, Clement C and Barka EA (2005b) Endophytic colonization of *Vitis vinifera* L. by plant growth-promoting bacterium *Burkholderia* sp. strain PsJN. *Appl Environ Micobiol,* 71(4):1685–1693.

El-Tarabily KA, Hardy GEJ, Sivasithamparam K, Hussein AM and Kurtboke DI (1997) The potential for the biological control of cavity-spot disease of carrots, caused by *Pythium coloratum*, by streptomycete and nonstreptomycete actinomycetes. *New Phytologist*, 137:495-507.

Glick BR, Penrose DM and Li J (1998) A model for the lowering of plant ethylene concentration by plant growth-promoting bacteria. *J Theor Biol*, 190:63-68.

Li B, Xie G and Soad AC (2005) Suppression of *Meloidogyne javanica* by antagonistic and plant growth-promoting rhizobacteria. *J Zhejiang Univ SCI*, 6B(6):496 496-501.

Lucy M, Reed E and Glick BR (2004) Applications of free living plant growth-promoting rhizobacteria. *Antonie Van Leeuwenhoek*, 86(1):1-25.

Marschner H (1995) Rhizosphere pH effects on phosphorus nutrition. In: Genetic Manipulation of Crop Plants to Enhance Integrated Nutrient Management in Cropping Systems (eds) Johansen C, Lee KK, Sharma KK, Subbaro GV and Kueneman EA, 1. Phosphorus. Proceedings of an FAO/ICRISAT Expert Consultary Workshop. ICRISAT Asia Center, India. Patancheru, Andhra Pradesh, India: International Crops Research Institute for the Semi-Arid Tropics. pp:107-115.

Ping L and Boland W (2004) Signals from the underground: bacterial volatiles promote growth in Arabidopsis. *Trends Plant Scis*, 9(6):263-266.

Rodrýguez H and Fraga R (1999) Phosphate solubilizing bacteria and their role in plant growth promotion. *Biotechnol Adv*, 17:319-339.

Ryu CM, Farag MA, Hu CH, Reddy MS, Wei HX and Pare PW (2003) Bacterial volatiles promote growth in *Arabidopsis*. PNAS 100:4927-4932.

Scherer HW (2001) Sulphur in crop production. *European J Agronomy*, 14: 81-111.

Steenhoudt O and Vanderleyden J (2000) *Azospirillum*, a free-living nitrogen-fixing bacterium closely associated with grasses: genetic, biochemical and ecological aspects. *FEMS Microbiol Rev*, 24:487–506.

Timmusk S and Wagner GH (1999) The plant-growth-promoting rhizobacterium *Paenibacillus polymyxa* induces changes in *Arabidopsis thaliana* gene expression: A possible connection between biotic and abiotic stress responses. *MPMI,* 12(11):951-959.

Valois D, Fayad K, Barbasubiye T, Garon M, Dery C, Brzezinski R and Beaulieu C (1996) Glucanolytic actinomycetes antagonistic to *Phytophthora fragariae* var. *rubi,* the casaul agent of raspberry root rot. *Appl Environ Microbiol,* 62:1630-1635.

Whipps JM (2001) Microbial interactions and biocontrol in the rhizosphere. *J Exptl Bot,* 52:487-511.

Woitke M and Schitzler WH (2005) Biotic stress relief on plants on plants in hydroponic systems. *Acta Hort (ISHS),* 697:557-565. Cited online from website http://www.actahort.org/books/697/697_73.htm

Mind Bogglers

1. What are the major mechanisms of spectrum of modus operandi of plant growth promoting action of PGPRs?
2. How plants communicate with these PGP micro-organisms? Is there any specificity of either plant host or microbe during the crosstalks?
3. How and why PGPRs having ACC deaminase are more benefited than the ones that lack it?
4. Comment on the statement that roots of micropropagated plants exist in a microbiological vacuum.
5. What type of resistance or immune response is initiated in plants inoculated with PGP microbes? Give some suitable examples of the type of diseases and causative phytopathogen controlled by inoculation of these microbes.
6. Discuss in detail how one can increase the establishment of inoculated PGPRs in the rhizosphere/endorhizopheric zones?

8

Mass Production and Quality Control of Biofertilizers

- Introduction 202
- Manufacturing of Biofertilizers 202
- Mass Production of Biofertilizers 206
- Methods of Biofertilizers Innoculation 208
- Current Scenario of Biofertilizer Production in India 211
- Quality Control of Biofertilizer 212
- Future Perspectives 221
- References 221
- Mind Bogglers 222

INTRODUCTION

India being a developing nation is trying to reach a balance in ever increasing population with marginal increase in crop production. There occurred a bigbang in crop production during the green revolution era however this also resulted in creation of infertile, barren lands with low chemical nutrient input responsive soils. Soil fertility is the major constraint among resource poor Indian farmers as lesser benefits are harnessed by use of newer agricultural protocols/tools/products *viz* improved varieties, productive technology from less fertile soils. There is dire need for looking for the alternative technologies of food or crop production to satiate the hunger of burgeoning Indian population. Today there is enormous emphasis on the conservation facets of soil fertility and crop productivity which are the instrumental philosophies in sustainable agriculture. Biofertilizers are the cost effective, ecofriendly and renewable resource of plant nutrients that could well supplement the use of already functional and popular chemical fertilizers for obtaining increase in crop yield without over hazardous impacts on various facets of soil chemistry, fertility and microbial diversity to attain sustainable agricultural system.

MANUFACTURING OF BIOFERTILIZERS

Major biofertilizers being commercialized till date include the nitrogen fixing *Azotobacter, Azospirillum* and *Rhizobium* strains for cereals and legume crops while phosphate solubilizing bacteria/fungi *viz Bacillus/Aspergillus*, plant growth promoting bacteria *viz Pseudomonas fluorescens* and biocontrol agents as *Trichoderma sp*. The manufacturing of bioinoculants requires following major steps that have to be followed sternly to ensure the production of a quality product (Gosal *et al*, 2001).

a) Strain selection: This forms the foremost criteria before production of any biofertilizer and generally also involves the strain improvement techniques (in case there is a repository of previously isolated cultures). The efficient strain selection criteria includes the isolation and characterization of competitive isolate with the ability to outgrow in the rhizosphere among the native and physiologically more versatile microbes and to get establish in large number in the rhizosphere or the region of colonization in host plant roots resulting in completion of the ardent enzyme complex or product for which the inoculant could be developed. Thus it involves exploring the biological potentialities of the strain for the respective crop and helps in screening out the best performing microbial form/ strain/ isolate. The basic research is required to be carried out such as deciphering the pH, temperature ranges as well as the agitation, aeration and seed inoculum rates. This primary work could help in delineating the optimum conditions for various criteria that could be used for large scale production of the inoculant.

Moreover laboratory *in vitro* systems, pot culture and field evaluation are prerequisite before launching of any biofertilizer to fully ascertain its efficacy as there are chances that the promising lab strains under aseptic controlled conditions may prove to be ineffective in field trials under natural conditions. This may occur due to several reasons as soil pH and fertility levels (soil edaphic factors), variations in competitive behaviour or ability to combat antagnostic or predatory microflora, host specificity/ range (as in case of rhizobial inoculants) and nitrogen stress invoked by the host plant. The selected efficient strain must possess ability to grow fast once inoculated and also should retain the desirable potentialities of the stable high number biological activity/ activities.

However the already known repository of efficient strains could be utilized to further improve the efficiency, biological activity or range of the biological activities of a particular strain. This could be performed by using genetic engineering/ molecular biological approaches that may include various protocols as inclusion/ transfer of novel plasmid/ plasmid borne genes by electroporation, transformation or conjugation.

Desirable properties/characteristics of strain in large scale biofertilizer production

- Ability to multiply under broth culture conditions
- Survival in carrier material
- Stability during storage
- Ability to withstand adverse conditions (environmental stress, agrochemical inputs) and to perform desired action under natural/ field conditions
- Competitiveness in terms of enhanced/early colonization of soil/ rhizosphere and internal plant organs as per the mode of action
- Enhanced establishment in a wide range of host genotypes

b) Preparation of inoculum culture: The preparation of the mother or the seed culture is done by growing a loopful of the bacterial/fungal growth in the respective specific culture broth medium (as yeast extract mannitol broth for *Rhizobium*, Jensen's broth for *Azotobacter*, malic acid Doberniers medium for *Azospirillum*, Pikovskaya's broth medium for phosphate solubilizing microorganisms) at 28-30°C in rotary shaker for 2 to 7 days.

c) Carrier materials and their processing: Three basic forms of commercial inocula are solid, liquid and freeze-dried. Since majority of commercially available biofertilizers include the carrier based inoculants the carrier material having the required properties for sustenance and survivality of bacteria/fungi and high water absorbance capacity is utilized. The basic carrier material used is peat/lignite or the activated charcoal though other alternatives have been tried on as saw dust, FYM, sterilized soil, ash, but lesser positive reports could support for their commercial use.

The most popular used ones are solid, peat-based inoculants that could be purchased for seed or direct soil application. Other alternative carriers include the vermiculite having multilamellate structure, which provides better aeration and higher water holding capacity and also higher adherence to the inoculated seed surface (Bisoyi and Majumdar, 2000). Liquid inoculants are available in broth culture or as frozen concentrate. Broth or frozen concentrates usually are mixed with water and sprayed into the seed furrow at planting. Being in liquid state the major constraint is that of storage or shipment as these inoculants must be kept frozen or refrigerated which further limits their availability through normal distribution channels. Bottle or agar slants, lyophilized culture and granular formulations (clay, alginate, polymer gels as polyacrlamide) are other types of solid base biofertilizers that have also been produced and commercialized by several foreign countries as USA, Australia and New Zealand (particularly liquid based formulations).

The unsterile carrier material contains several types of contaminates (physical and biological) that could decrease the shelf life of the inoculant under packaged conditions during distribution and storage. Thus to avoid substandard biofertilizer production pretreatments of the carrier such as sterilization by autoclaving/gamma irradiation should be performed.

d) Packaging, storage and distribution: The inoculum is mixed with the carrier (in case of carrier based formulation) thoroughly i.e. blending and this is packed and sealed in gamma sterilized polythene bags. Blending is generally performed as manual mixing or by use of a mechanical mixer followed by curing of the mixed inoculum for 24h that help in acclimatization of the inoculum with the carrier.

The cured carrier based inoculum is packed in low density thick polythene bags (200 microns) to avoid moisture loss during distribution and storage. Each bag is filled with equal quantity of inoculant (generally 200 to 250g) and then sealed using an electric sealer. The packets always have the instructions for use/application, date of manufacture/expiry, batch number and name of the inoculant and beneficiary crop(s). However the liquid based formulations do not require any mixing with the carrier but these may involve additions of sunscreens, adjuvants or stabilizers to sustain high microbial load and negligible to very low number of contaminants. Granular formulations are prepared by mixing the inocula with clay, alginate polymer gel or any other granular component.

The packed inocula are incubated for a week at room temperature to obtain microbial multiplication and growth that reaches to a required standard followed by storage at cold temperature (4^0C to 15^0C).

e) Factors affecting inoculant survival in carrier: Several biotic and abiotic factors are responsible for survival of microbial inoculant in the carrier material such as level of contaminants (number/count) for the former and pH, temperature, aeration, substrate concentration and moisture content for the latter factor respectively.

MASS PRODUCTION OF BIOFERTILIZERS

Mass production of biofertilizers requires certain prior research on efficient strain, method of mass production (aerobic flask shake or fermentation technique) and carrier material (required to be cheap and suitable for long term storage) to be utilized for large scale inoculant production as well as are foremost for acceptance of the bioinoculant by the end users/ farmers.

Shake flask based inoculant production: The shake flask culture production could be useful for small scale biofertilizer producers. This type of production system involves use of 500ml culture bottles/flasks making it vulnerable/difficult to standardize or optimize the whole system. However it depends largely on inclusion of manual labour for steps as inoculation, blending, packing and sealing. It is an aerobic cultivation of microbes in flasks at optimum temperature and agitation rates on a rotary shaker for 2-4 days depending on the type of microbe. The culture so obtained is mixed manually with carrier and packed in polythene bags.

Fermenter based inoculant production: The fermenter is automatic and accurate method of multiplication of any microorganism and performs both media sterilization as well as microbial multiplication. Fermenter based inoculant production requires more sophisticated techniques, skillful technicians to handle and financially more inputs. It is easy to optimize whole production unit than the former technique. The fermenter based inoculant production includes use of commercial or industrial fermenters (50 to300L capacity) for aerobic production of several biofertlizers as *Rhizobium, Azotobacter, Azospirillum, Pseudomonas, Bacillus* etc to obtain higher microbial load (10^8 to 10^9) in the final product.

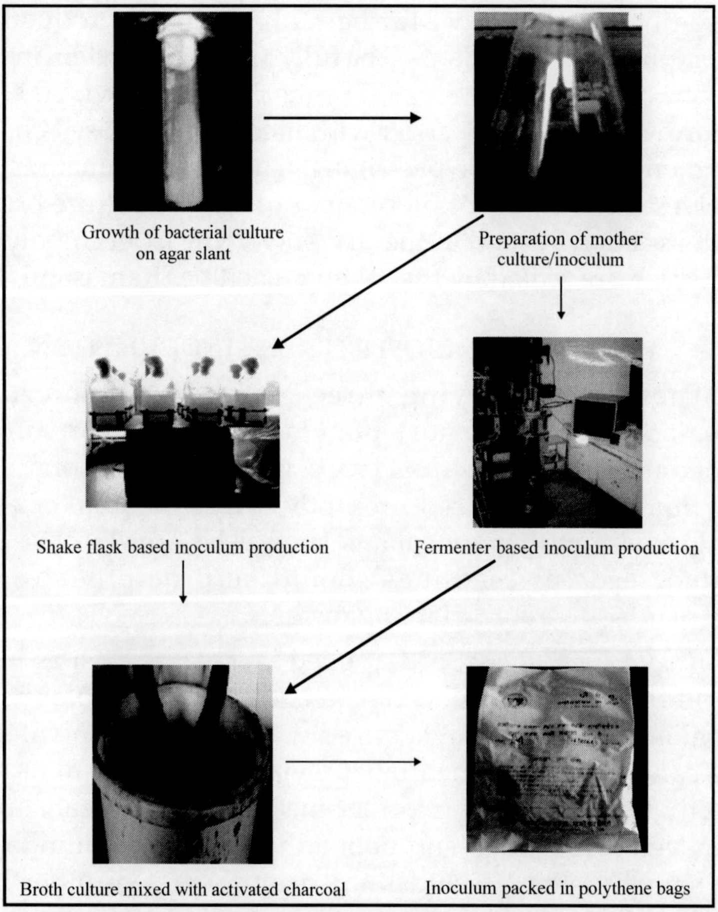

Flow diagram showing steps in mass production of biofertilizers

(For a coloured version of this figure, see plate section, page 303)

The major steps involved during fermenter based bioinoculant production include media sterilization, adjustment of the pH of the media, media inoculation by 24h old live culture @ 1 per cent inoculant level followed by provision of air (blown in media using pump) at optimum agitation rate favouring bacterial multiplication. As this production technology involves the use of sealed fermenter, the aseptic conditions could be maintained up till the production and packaging stages. Newer instruments as mass flow dispensor, microbial growth sensors & modern automation including use of semi- or fully automatic packaging systems have further enhanced the production of better quality bioinoculants. The fully automatic system involves use of Form, Fill and Seal (FFS) machines (Bisoyi, 2000) that perform mixing of the carrier with inoculum in closed chamber and then collection for packaging. The semiautomatic system involves the delivery of preset quantity of the culture broth by broth injection system in the dry powered carrier in polythene bag which are sealed by the sealing machine (Ramasami, 2000).

METHODS OF BIOFERTILIZER INOCULATION

There are following three methods of biofertilizer inoculation that are most popularly utilized for different agronomic and horticultural crops. In general the biofertilizer/ inoculum should be used promptly before the date of expiry. Being living entities inoculants should be properly used and treated seeds or ceedlings should be promptly sowed or transplanted to avoid viability losses.

i) Seed application: Seed-applied inoculants exist as planter box additives, preinoculated seed and custom inoculants. Most commonly the inoculant is mixed with seed in the planter box. This can be accomplished by applying dry inoculum or slurry directly to seed with the former method least desirable owing to uneven distribution and poor adhesion of inoculum to seed. The slurry method includes the use of water or other liquid solutions that make inoculum to better adhere to the seed coat. Seed may also be pre-wetted before mixing it with dry inoculants.

Treating seeds with carrier based biofertilizer/inoculant

These carrier based cultures are mixed with minimum amount of water or with sugar/gur/pharmaceutical grade gum arabic (10%) solution to form slurry and are applied on the surface of seeds as a uniform coating with the later added to enhance the survival rate and multiplication of inoculated microbial form. The seed applied with slurry are spread in shade to be dried before sowing and are immediately sown after shade drying.

However other techniques as the use of preinoculated seeds by addition of carboxymethyl cellulose or gum arabic as pelleting material enhances the survival of the inoculant cells under harsh soil edaphic conditions (as saline or sodic pH values of respective soils) or other agronomical procedures as fertilizer or pesticide application. The seeds are treated with the above inoculant along with ground gypsum/calcium carbonate leading to coating of the seeds that could be sown after shade drying.

A sticker can significantly increase the adhesion of inoculant to the seed surface. Without a sticker, the product will fail to adhere to the seed, and accumulate in the bottom of the planter box. The most convenient sticker is one that is integral or actually a part of the product, allowing a quick, clean and easy inoculation.

ii) Seedling inoculations: This technique is used in transplantable crops such as onion, rice, tomato, brinjal, chilly and many horticultural or floricultural crops and involves dipping of roots of the seedlings (one month old) in inoculant (liquid or carrier based) for 30 to 45 mins followed by their transplantation. In general *Azospirillum, Azotobacter* and PGPR treatments are performed by this method. Two to 2.5 kg inoculant (carrier based) is used for inoculation of seedlings required for one acre of land.

iii) Soil treatment: This could be performed for both short and long duration crops and involves preparation of a mixture of biofertilizer (3-4kg for the former and 4-5kg for the latter) in soil/compost (90-120kg short duration and 120-150kg long duration) which is broadcasted in a hectare of land either at last ploughing or in the standing crop just before irrigation. Primarily the soil application of biofertilizer is performed for sugarcane, fruit crops and trees where 20 to 30g of biofertilizer is added along with compost or farmyard manure per sapling while in sugarcane 5 to 6kg biofertilizer is added to soil or compost for one acre before earthing i.e. after two to three months of planting.

Seedling inoculation before transplantation

CURRENT SCENARIO OF BIOFERTILIZER PRODUCTION IN INDIA

The potential of biofertilizer use has been estimated by various groups state wise and agroclimatic region wise. The basis of such estimate involve area sown under different crops and the rate of biofertilizer used for seed treatment while biofertilizers that are applied in soil the rate of soil application could be taken for calculation of the potential.

Estimates of biofertilizer production by different agencies in India

Strain	Area sown ('000ha)	Dem and under 100% area (MT)	Estimated by NBDC (MT)	Estimated by BCIL (MT)
Rhizobium	62,257	32,585	34,999	19,000
Azotobacter	49,920	56,632	1,45,953	11,400
Azospirillum	51,553	43,800	74,342	54,400
PSM	1,63,730	1,33,019	-	-
BGA	-	-	2,51,738	260,000
Total	**3,27,460**	**2,66,036**	**5,07,032**	**3,44,800**

Source: Report of task force on biofertilizers (May 2001), Department of Fertilizers, Ministry of Chemicals and Fertilizers, New Delhi

There is an increasing trend of use of biofertilizers due to popularity of cultivation visions as organic farming and sustainable agriculture. The demand thus increased would escalate the production and marketing of biofertilizers. The present annual production of biofertilizers in India is approximately around 10,000 tonnes (Bagyaraj, 2004) while the present annual requirement is 2.5 times the total production (around 25,000 tonnes/annum).

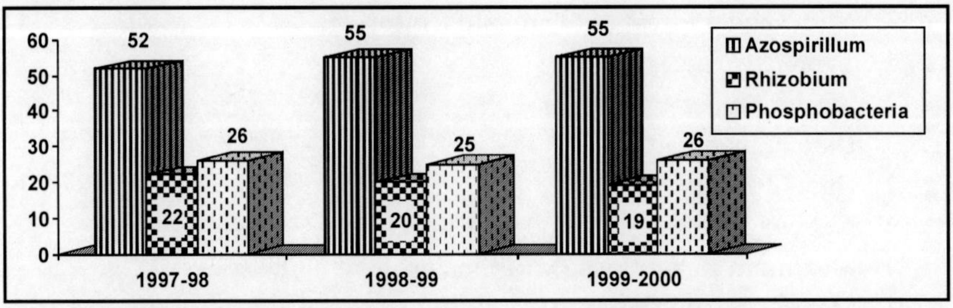

Production pattern and per cent share of popular biofertilizers produced during 1997-2000

Quality Control of Biofertilizers

Unpredictability of performance is the characteristic common to most of biofertilizers which requires research and reviewing to enhance the consistency in performance and parameters as competitiveness and rapid/sustained establishment of host endorhizosphere or rhizosphere soil for increased benefits. The other constraint is the presence of very low counts of indigenous beneficial microflora and benefits in terms of enhanced crop yield could only be obtained by continued use of biofertilizers consecutivelt for three to four years. Biofertilizer inoculation in crops does not exhibit dramatically overwhelming results as compared to use of chemical fertilizers. Thus continued inoculations are required that help raise the viable competitive bacterial counts resulting in enhanced establishment. Biofertilizers could not provide overall net financial benefits within same year or two of their inoculation. The prime thing required is to reach a certain standard of the microbial count i.e. a threshold level of population that later helps in obtaining benefits in terms of high yield compared to the uninoculated fields/plots. This viable count of live beneficial desired microbes is the criteria for the functional quality of the bioinoculant (product) and any tempering with the product contents or the specifications makes suspicions regarding the quality.

Details required be printed on biofertilizer packet (each)

- Name of the product/name of strain
- Crop/crops for which it is recommended
- Name and address of the manufacturer
- Type of carrier
- Batch number
- Date of manufacture
- Date of expiry
- Net weight of product and area meant for
- Maximum Retail Price (MRP) inclusive of all taxes
- Method of application
- Storage instructions

Importance of quality control: Indian farmers are not very welcoming and fervent to adopt this facet of sustainable agriculture because of the poor quality of the bioinoculants produced and supplied to the farmers which may result in strengthening the wave of incompetency of the biofertilizer technology as such. Though there are specifications delineated by BIS (BIS standards) for four species *viz. Rhizobium* (IS:8268-2001) and *Azotobacter* (IS:9138-2002), *Azospirillum* (IS:14806:2000) and PSB (IS:14807:2000), there is no systematic quality certification system and monitoring mechanism as quality testings are not mandatory for the producers which makes it an internal arrangement and voluntary system. Majority biofertilizer production units lack inhouse quality testing facilities i.e. lab infrastructure and plans/arrangements which are prerequisite for these products (involve living microorganisms). Regular quality check ups and batch-wise certifications are highly essential that help in delineating the quality of the product best before use. BIS standards include quality tests regarding cell number not to be $<10^7$ per gram of carrier at the time of manufacture and specification of cell count and permissible contamination at expiry dates. However the BIS specifications for the liquid biofertilizers are not being delineated and this aspect is required to be addressed since the present trend showcases escalated interest in production of liquid biofertilizers that are proposed to have higher cell count as well as shelf life owing to less or negligible contamination. Infrastructural facilities as creation of accredited labs for testing the quality of biofertilizers are required along with the legislation for quality monitoring and ensuring marketing and supply of good quality biofertilizers to their end users. Each unit therefore should have the following facilities:

- adequate microbiological laboratory and qualified microbiologist
- Sampling and testing at various stages of production, including the quality of raw materials

- Specifications on the packet regarding the contents and cell counts.
- Fixing of quality certificate and batch number and packing of the products in proper packing material
- products should be stored in cooler places till their distribution/selling to farmers
- aseptic conditions have to be ensured (cleanliness and contamination free production lines and housing systems) to avoid contamination of inoculant
- automatic and closed systems are recommended

Methods of quality control: Quality control of biofertilizers involves evaluation of inoculant quality by enumerating the viable cell count of the inoculant. The quality control tests include both conventional (MPN counts and infection tests for *Rhizobium* inoculants, serological protocols as ELISA, Immunoblot, FA test, agglutination etc) as well as unconventional techniques (RFLP, PCR based techniques, reporter gene techniques). However among the conventional diagnostic tests serological tests are the most rapid ones for the biofertilizer production units while the MPN count is time consuming and unreliable method of assurance. The on-field quality assessment could be performed by use of reporter genes as *gus, lacz, gfp* etc which produce color change on specific substrate or under particular radiation, for quality assurance of the bioinoculant. As per ISI standards, one gram of biofertilizer immediately after its preparation should have one crore cells of bacteria and 15 days before expiry date one gram of biofertilizer should have 10 lakh bacteria.

Constraints of biofertilizer production and use

Being a biological entity, biofertilizers are subjected to several constraints stationed at various aspects of biofertilizer production and commercialization and primarily include biological, technical and marketing constraints (Bagyaraj, 2004).

a) Biological and technical constraints

- Overall the biological constraints include stability, competitiveness and stress encountered by the inoculated microbes due to ecological factors encompassing both biotic as well as abiotic factors. Being a biological entity, biofertilizers/inoculants are prone to inconsistent performance in fields as many soil microflora (bacteriophages, myxobacteria, protozoans, cilliates) and fauna (nematodes, insect larvae and earthworms) are predators that devour upon the inoculated microbial forms thereby decreasing their number and survivality in soil and thus affect positive benefits so imparted by biofertilizer application. The abiotic factors that affect the biofertilizer efficiency include the hiked temperatures encountered during summer seasons, soil moisture content, soil pH i.e. acidity/alkalinity and macro as well as micronutrient status.

- Among the technical constraints encountered by the biofertilizer production following are prime

  • It is prerequisite that certain biofertilizers are crop specific particularly legume crops require host specific strains while other bioinoculants have to be region/soil specific. This specificity criteria has to be fulfilled as changes in soil/region/crop directly alter/affect the microbial interaction with the inoculated plant as well as with the surrounding rhizospheric abiotic and biotic agencies/factors.

  • The most common/popular carrier material used in biofertilizer production is peat/lignite and this being available at few regions (Southern Nilgiris and Neyveli regions) in India result in increase in the total cost of production as it has to be transported to the manufacturing unit. The transportation charges thus exceed the cost of production, which thereby debars the popularization of the product and easy approach to the marginal and poor farmers who are the actual buyers of this technology. The type of carrier material (peat, lignite, activated charcoal, vermiculite, saw dust, autoclaved soil, paddy husk, crushed groundnut shell, farmyard manure, compost, pressmud with charcoal, Indian peat soil etc) affect the viability of microbial cells. Saha *et al* (2001) have observed the comparative viability of *gfp* transformed *Rhizobium*, in four different bioinoculant carriers *viz* lignite, paddy husk, sawdust and crushed groundnut shells at various temperatures (28°C, room temperature, 37°C and 40°C) and reported maximum viable count of *Rhizobium* transformant cells in lignite while minimum in groundnut shell carrier at all temperatures. Moreover maximum protection was imparted to the rhizobial cells by paddy husk in comparison to sawdust and crushed groundnut shells at all temperatures but the maximum number of cells remained in paddy husk at 28°C.

- The final product particularly the carrier based bioinoculants have a shorter shelf life period making the product vulnerable for use after or near to its expiry date. Though the shelf life period of the liquid biofertilizers has been extended to one year still these formulations require storage and distribution at refrigeration temperatures (4°C).

- Substandard quality of inoculants is another technical constraint that is the result of use of poor quality of manufacturing material/conditions as well as poor storage facilities.

- There is also a prime constraint of lack of or absence of well equipped quality testing laboratories to ascertain the quality of the product and the strict follow up of the implementation of mandatory certification criteria by the manufacturing unit.

b) Infrastructural Constraints

The infrastructural constraints include inadequacies regarding space and electric supply in the building used as the biofertilizer production unit.

c) Marketing constraints

Many marketing constraints exist in wide spread use of biofertilizers and these constraints have been due to poor marketing network and weaknesses in the retail outlet and storage facilities

- most of the biofertilizers are required in particular season with special reference to the legume bioinoculants and thus there is only seasonal demand of a particular biofertilizer in a particular season. The dealers should understand the shelf life and crop specific use of various biofertilizers as this knowledge would be useful for planning out of the sales accordingly.
- biofertilizers are in limiting demand because of farmer's unawareness about biofertilizers and hence farmers have to be educated regarding the capability and proper use of biofertilizers. Uniform packaging of 250g (seed treatment) and 1kg (soil treatment) is required to avoid confusion in farmer's mind.
- there is lack of perception of concerned official(s) about the importance of biofertilizers from sustainability point of view. The State and district extension officers (Agriculture) are themselves not much informed and well versed with the production technology and benefits of biofertilizers and thus should be educated about various aspects of biofertilizer technology
- the lack of qualified technical staff and mishandling during storage and transportation culminates to substandard quality of product

Factors affecting effectiveness of biofertilizers: The critical factors which are responsible for the effectiveness of a particular bio-fertilizer are as follows:

- Suitability of the species to the target crop (as *Rhizobium* biofertilizers for legumes)
- Strain specificity for various soil environmental (agroecosystem, soil edaphic factors) conditions

- ❒ Maintenance of aseptic manufacturing conditions to restrain contamination and to obtain high living cell count of pure inoculant in carrier material.
- ❒ Production qualities including quality of the packing material, which determines the shelf life
- ❒ Poor distribution and storage conditions drastically affect product quality
- ❒ Biofertilizer application techniques differ according to the type of crop and could affect the benefits thus harnessed from the crop
- ❒ Soil conditions particularly pH, organic matter content and moisture level and agronomic practices.

The exact potential of biofertilizers is not being harnessed owing to lack of knowledge among farmers and extension workers regarding inoculation technology and the kind of host plants that could be most benefited by inoculation of a particular biofertilizer. The farmers have to be educated and encouraged for attaining a level of scientific temper and methodological use of few techniques for better production. The farmers should be persuaded to test their soil type/soil nutrient status, plant and cultivar types to obtain background information required for obtaining benefits in terms of hiked yield and production. Earnest measures are required for formulation of effective inoculant delivery or supply system.

- ❒ To use efficient or competitive, region/crop specific strains
- ❒ To use good quality sterilized carrier material
- ❒ To employ technically competent human resources for production, storage and marketing. The farmers are needed to be properly educated and convinced about this important technology through wide publicity and demonstration trials. Biofertilizers must be marketed as other FMCG products and the work on biofertilizers in SAUs is required to be supported by financial help from private sector in form of sponsors.
- ❒ To ensure production of good quality biofertilizers having ISI standards so as to instill confidence in farmers
- ❒ To establish nodal agency for quality monitoring during production, transport, storage and use and also it is advisable if these biofertilizers got tested/recommended by State Agricultural Universities

Research gaps required to be worked out: Apart from other constraints certain research gaps have to be worked out before plunging for mass production and marketing of biofertilizers such as

- To evolve competitive and crop/region specific through selection/ genetic manipulation
- To carry out research regarding integrated use of various biofertilizers

VAM biofertilizers : Constraints regarding production and usage

- obligate symbiont
- cannot be mass multiplied in pure culture by any known method
- usually maintained and multiplied in pot cultures which are both bulky and also donot maintain the purity of VAM cultures
- VAM based biofertilizer currently recommended only for transplanted and nursery raised crops.
- *In vitro* culture methods are alternative for maintenance and multiplication
- Results depend on plant available P status of soil and same regarding yield of inoculated plants

Nitrogen fixing biofertilizers : Constraints

- Mostly host specific as rhizobial biofertilizers
- Benefits differ by use of fungicides or compounds toxic to inoculum
- shelf life aspect of these biofertilizers has to be improved. Though new alternatives are being rushed in the biofertilizer market including the liquid biofertilizers that contain glycerol or other adjuvants to increase the shelf life around a year, the efficacy of these concomitants is still to be adjudged.
- Results of *Azotobacter/Azospirillum* biofertilizer application on yield vary due to process of loss of fixed nitrogen by an array of denitrification reactions performed by various denitrifying soil bacteria.
- benefits of diazotroph inoculation decreases over a period of time due to decreased soil count owing to competition in rhizosphere with the indigenous microflora primarily the bacilli and pseudomonads.
- cyanobacteria are not much popularized in Punjab and northern states because of depletion of inoculant strain by fluctuating hiked temperature, presence of protozoan predators

- *Azotobacter* inevitably requires presence of high content of soil organic matter before inoculation limits its use and popularity. Punjab soils have become deficient in soil organic matter due to continuous and intensive monocropping practices and thus the inoculated bacteria could not survive as the organic acids present in matter are utilized as substrate by the bacteria. Moreover nitrogen fixed by *Azotobacter* could only be released upon cell death.

Government Intervention in Biofertilizer Market

To attain production targets, the Government of India implemented a central sector scheme called National Project on Development and use of Biofertilizers (NPDB) during the Ninth Plan for the production, distribution and promotion of biofertilizers. A National Biofertilizer Development Centre was established at Ghaziabad as a subordinate office of the Department of Agriculture and Cooperation with six regional centers. The purpose of the scheme covered organization of training courses for extension workers and field demonstrations and providing quality control services. Production and distribution of different biofertilizers were also undertaken but subsequently discontinued as the centers redefined their role towards R&D and HRD related activities. Capacity creation and production was however encouraged through one time grant for new units.

The public sector organizations form a bulk of the units in the industry, while similar units in the private sector are also coming forward. Different State governments also provide subsidies sometimes up to 50% of the sales realization though through unsystematic subsidization which may lead to discrimination and manipulation and thus a lot of intra industry variation in prices.

The government also plays a dominant part in marketing biofertilizers in three possible channels:

(a) State government *via* District level Officers and Village level workers to farmers, *(b)* State Marketing federation via cooperative bodies to farmers and *(c)* State Agro-industries

FUTURE PERSPECTIVES

Today's farmers are imbibing new technology for sustainable agriculture and the very cult of organic farming has enhanced the interest of farmers in biofertilizer use and benefits associated to their application. New methods for preparation of biofertilizers as alternatives of the carrier based cultures are being developed and these would revolutionize the acceptance and wide spread use of microbial inoculants for increased yield. New techniques particularly molecular methods have been employed now for obtaining assurance regarding the presence and establishment of bioinoculants in various soil and plant niches. These methods involve use of reporter/marker genes and have been instrumental for providing a qualitative as well as quantitative assessment of the benefits reaped by inoculation of biofertilizer in a particular crop.

References

Bagyaraj DJ (2004) Quality control and constraints in biofertilizer production technology. *In: Biofertilizers Technology for Rice based Cropping Systems*, Kannaiyan S, Kumar K and Govindarajan K (eds), Scientific Publishers, Jodhpur, India, pp. 402-408.

Bisoyi RN (2000) Constraints of biofertilizer production and remedial measures. *In: Ibid*, pp. 50-54

Bisoyi RN and Majumdar P (2000) Technology for increasing shelf life of biofertilizers. *In:Ibid*, pp. 44-49.

Gosal SK, Gupta RP, Pandher MS and Kalia A (2001) Biofertilizers - potentials and constraints. In : *Applied Microbiology*. Trivedi PC (ed), Agrobios India, Jodhpur, pp. 1-10.

Ramasamy S (2000) Automated mechanized systems in biofertilizer production:MFLs Experience. *In: Proceedings of Symposium on Current Status of Biofertilizer Technology and its Promotion in Karnataka, Tamil Nadu, Kerala and Pondicherry,* R.N.Bisoyi (ed), RBDC, Banglore, pp. 35-37

Saha AK, Deshpande MV and Kapadnis BP (2001) Studies on survival of *Rhizobium* in the carriers at different temperatures using green fluorescent protein marker. *Curr Sci*, 80(5):669-671.

Mind Bogglers

1. Give a brief note on the desirable features for a better bioinoculants.
2. Enlist various steps involved in mass production of biofertilizers. What would be the preferred technique for mass production for small scale unit?
3. Quality control of biofertilizers is essential. Comment on this statement.
4. What are the quality control criteria for ascertaining the quality of *Rhizobium* and *Azotobacter* inoculants?
5. What are the various types of biofertilizers depending on the consistency of the final product? What are the advantages and disadvantages of use of liquid biofertilizers?
6. "Rich soil microbial diversity is the index of agricultural prosperity of the nation". How this microbial diversity could be maintained? Give your view point on the role of biofertilizers for sustainable agriculture.
7. What are the constraints associated with bioinoculant production, marketing and utilization?

9

Biopesticides

- ❐ Introduction 224
- ❐ Bioinsecticides 226
- ❐ Bioherbicides 238
- ❐ Biofungicides 241
- ❐ Future Perspectives 247
- ❐ Conclusions 250
- ❐ References 251
- ❐ Mind Bogglers 255

INTRODUCTION

Introduction of improved high yielding varieties of crop plants, use of modern agricultural practices coupled with essential control of pests of food and fiber crops have significantly contributed towards increasing the food production in green revolution era in India. Pesticide application forms the foremost limb in hiked agricultural food production and management practices. The production and post production management of crop yield requires increased use of pesticides.

About 80,000 tons of pesticides are used in agriculture in India annually (Srinivasan, 1997), mostly in cotton that accounts for 45% of pesticide usage (Dhaliwal and Pathak, 1993), rice (accounting for 23% of pesticide usage) and a significant proportion accounted for vegetables and fruits. However, in recent years increasing concerns have been expressed by various environmental groups on the inherent risks and problems involved in the indiscriminate and overuse of certain chemical insecticide products to obtain high yields, which can cause acute to chronic toxicity to non target organisms including man (Commandeur and Komen, 1992). Recent estimates of annual crop losses in India due to pests and diseases amount to around Rs. 2,90,000 million crores and one single insect *Helicoverpa* is responsible for losses of over Rs. 35,000 million crores per annum because of high pesticide residue content (Sharma, 2003). The incidences of pest resistance have aroused since last two decades with crop losses amounting to between 20-30% of total production. The sub-standard quality of pesticides having low concentration of recommended pesticides lead to chances for development of resistance pest generations and excessive indiscriminate use of pesticides have further putrefied the situation. This state of quagmire has culminated to suicidal attempts by farmers in mid- and southern India. There are several stern reports that suggest development of resistance to pesticides in important pests with its first incidences recorded for DDT and HCH in 1963. Today, major crop pests have been developing resistance towards several popular pesticides with one noteworthy example being the cotton bollworm resistance where the American bollworm having developed resistance to almost all pesticides in a number of regions, and particularly serious in parts of Punjab, Haryana, Andhra Pradesh, Karnataka and Maharashtra.

Moreover being more resistant to degradation by abiotic and biotic agencies the pesticide residues are observed to leach down to lower strata of soil, easily get absorbed by plant roots, accumulate in the food chain to be biomagnified in the food web. This feature of biomagnification of pesticides in plant and animal tissues (particularly lipid bodies) makes their use to be health hazardous as this may lead to several ailments. Indian food grains, agricultural and dairy products contain very high concentrations of pesticides. Therefore, to attain acceptable yield and quality of various crops, stress is now being laid on evolving alternate pest control measures. This ecological threat of pesticide usage is also increasing the interest for alternative pest management technologies to be pondered upon and applied in near future.

The alternative techniques to amend and decrease the impact of chemical pesticide usage form the foremost strategies of the Integrated Pest Management protocols. These include cultural methods for plant resistance to pests, conservation of natural enemies in the crop and the use of insect pest control products such as "Microbial Insecticides". IPM concept partly includes integrated use of biocontrol agents (microbes, insects, extracts/compounds) along with judicious, timely and less use of chemical pesticides as well as wise management of agricultural tools and cropping practices/protocols.

Types of Biopesticides, important part of biological control system, are products/processes derived from animals, plants and microorganisms such as bacteria and viruses. The most commonly used biopesticides include *Bacillus thuringiensis* (Bt) based products, Baculoviruses and botanical extracts as the use of rotenone, pyrethrin, nicotine, azadirachtin (alkaloid produced by neem). In addition to these, *Trichoderma*, a fungicide has been used proficiently along with biocontrol agents, such as *Trichogramma*, which are parasites and predators of pests and their *Bacillus thuringiensis* (Bt) modification or Bt-based products. Microbial biopesticides include microorganism(s) applied periodically to obtain biological control of target pest(s) in the applied field.

Approaches used for exploiting microorganisms for insect pest control

1. Colonization (Inoculative augmentation). It includes introduction of microorganism into pest population where it maintain and spreads itself.
2. Conservation (Environmental augmentation). It refers to enhancement in efficiency of pathogens, naturally occurring, by manipulating the cultural practices or other environmental conditions.
3. Microbial Insecticides (Inundative augmentation)-It involves dissemination of microorganisms in the pest population in large quantities in a manner similar to application of insecticides.

Botanical extracts (Nicotine, rotenone, neem)

- *Azadirachta indica* based products are non-carcinogenic to humans and non-toxic to birds and mammals
- several chemicals 'azadirachtin' most effective ingredient
- works as a broad spectrum insect growth regulator by disrupting molting during an insect's larval, or juvenile stages
- affects reproductive and digestive process of a number of important pests and also has strong anti-feedant properties
- commercially effective formulations of neem produced by few companies in India
- little of produce sold locally as most being for export markets

BIOINSECTICIDES

A large number of microbial species belonging to various groups have been found to be entomopathogens (Lacey *et al*, 2001), which, therefore, can be tried as biological control agents including entomopathogenic bacteria, viruses, fungi, protozoa and nematodes.

a) Entomopathogenic bacteria: Bacteria which are used for the control of insect pests are *Bacillus thuringiensis, Bacillus papilliae, Bacillus sphaericus, Pseudomonas fluorescens and Serratia entomophila.* Of all these, that have been evaluated, the most attractive by far has been the bacterial insect pathogen, *Bacillus thuringiensis*.

Fruits of *Azadirachta indica*

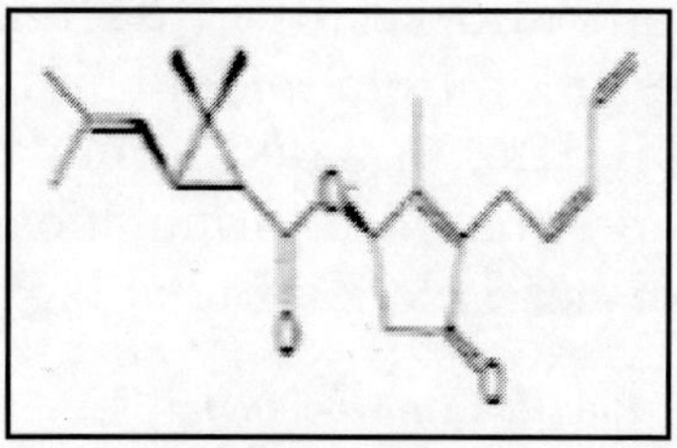

Azadirachtin (active compound with biopesticidal activity)

Chrysanthemum cinerarifolium

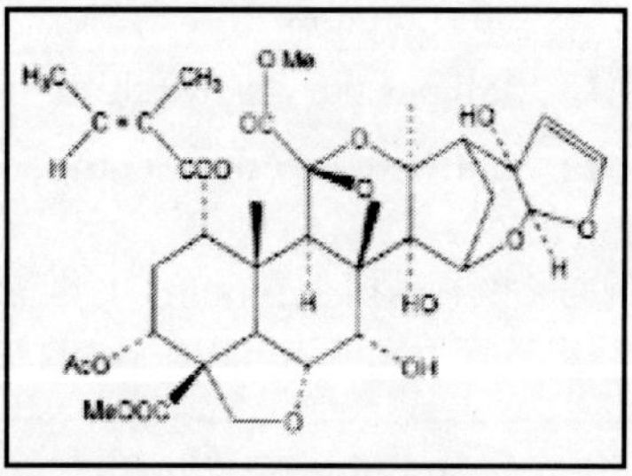

Pyrethrin

(For a coloured version of this figure, see plate section, page 304)

Bacillus thuringiensis: It is gram positive, aerobic, sporulating rod shaped bacterium that produces proteinaceous parasporal inclusions during sporulation including delta endotoxin and beta, alpha and gamma exotoxins which are highly insecticidal at very low concentrations. The specific toxicity of these crystalline inclusions against insect pests makes this organism a potential agent for biological control. There exists 34 recognized species of *B. thuringiensis* including most commonly used subspecies *kursataki, israelensis,* and *tenebrionis* active against lepidoptera, diptera (mosquitoes and blackflies) and *Leptinotarsa decemlineata* (Whalon and McGaughey, 1998). The crystal proteins are biodegradable and are specific to each subspecies/strain and are coded by a single plasmid borne gene. The crystal proteins could be classified on basis of amino acid sequence homology and presently fall under 40 different classes.

Among these classes some candidate proteins happen to exhibit specificity to multiple insect orders. These crystalline proteins have the basic target being the insect midgut columnar epithelial cells and result in decreased absorption of minerals and nutrition from midgut and finally death of the columnar cells. Till date, Bt based formulations (67 registered products with more than 450 formulations) occupy key position accounting for nearly 90% of total biopesticide sales (Neale, 1997).

Bacillus thuringiensis

- globally most commonly used biopesticide
- pathogen of lepidopterous pests (American bollworm in cotton and stem borers in rice)
- kill pest larvae by release of crystalline protein damaging insect midgut
- marketed by three companies in India. The total sale in 1999 was about 70 tons.
- Bt Product: Biobit, Dipel, Skeetal, Bactospeine, M-One, M-Trak Trident are commercially available

Ingestion *B.thuringensis* spores and crystalline protein

↓

Dissolution of ICP in alkaline midgut (pH 9.5)

↓

Activation by protease to produce active Delta-endotoxin (60 kDa)

↓

Irreversible binding of C-terminal domain of activated ICP to receptor

↓

Formation of pores and channels in gut membrane

↓

Equilibration of ions occurs

↓

Gut gets immobilized and epithelial cells lyse

↓

Insect stops feeding and gut pH is lowered by equilibration of blood pH

↓

Lethal septicemia and death

Mode of action of the Bt cry toxin (Schnepf *et al*, 1998)

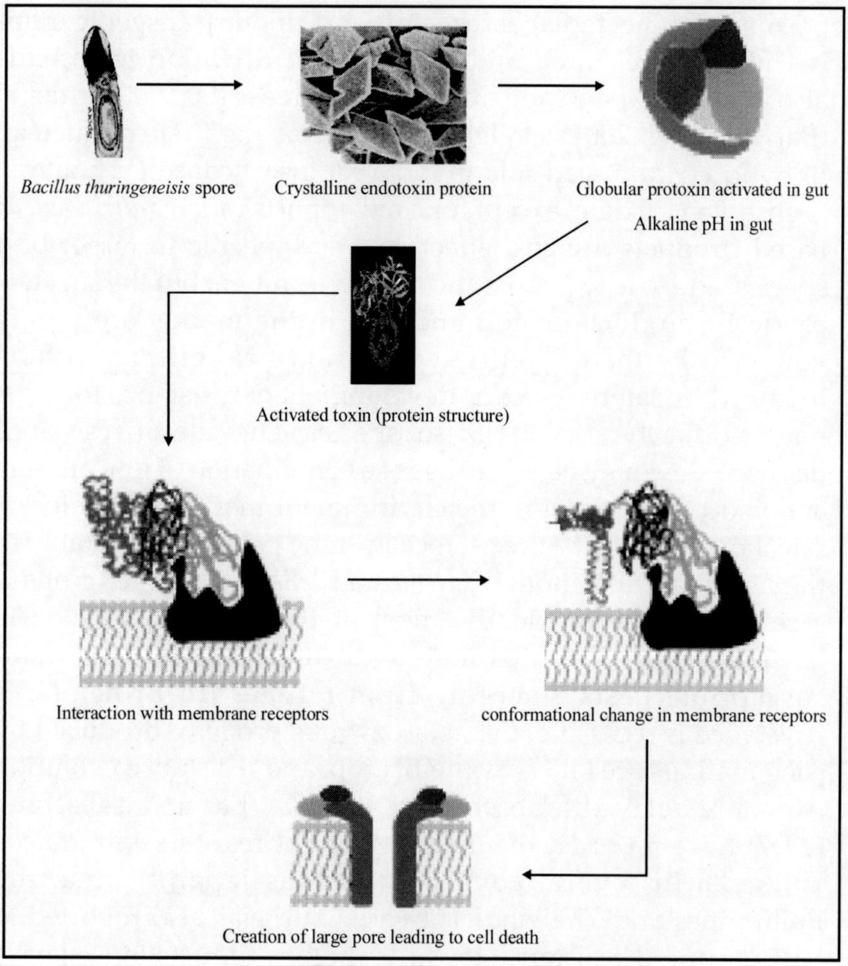

Bt Cry toxins : types

- CryI and CryII toxins : active against lepidopterans
- CryII and CryIV : active against dipterans,
- CryIII toxin : active against coleopterans

(For a coloured version of this figure, see plate section, page 305)

Application of Bt based formulations and field trials/results

By 2004, 17 countries have adopted a variety of transgenic crops including the Bt corn and cotton, herbicide glyphosate/ glufosinate tolerant corn, cotton, soybean and canola accounting around approximately 81 million hectares (Naranjo *et al*, 2005). Though the leading country that has maximum

(47.6 million hectares) acreage of land under transgenic crops is United States, area under Bt cotton cultivation in India is also likely to expand and expected to increase (500,000 hectares) (Barwale *et al,* 2004). Bollgard cotton was the first commercial Bt hybrid cotton seed sold in India for insect control. Though it is observed till date, except for a few reports that *B. thuringenisis* based products are cost-effective, very specific to target pest species, and safe to people and the environment but the survival of the Bt products or formulations in the market are purely governed by the cost effectiveness and field efficacy criteria followed by safety issues as development of resistance towards one or other type of the Bt subspecies. The rate of resistance has been slow may be due to use of a combination of low product and short persistence in the environment that limits the insect selection pressure but resistance in some pests, most notably in the Diamondback moth (*Plutella xylostella*) on crucifer crops is a serious concern. Ideally, most of the rare resistant pests emerging from Bt crops mate with the relatively abundant susceptible pests emerging from refuges. In principle, if resistance is recessive, the heterozygous progeny produced by such matings are killed by the Bt crop, and resistance evolution is greatly delayed (Tabashnik *et al,* 2005). Liu and Tabashnik (1997) have reported the development of resistance in *Plutella xylostella* in several countries such as Japan, China, the Philippines and Thailand. The same bug was also reported to have acquired resistance towards the Bt products in Malaysia (Iqbal *et al,* 1996).

b) Entomopathogenic viruses (Baculoviruses) : Baculoviruses among viruses have been gaining wide spread recognition and importance as possessing the ability to develop into potential insecticides (Mishra, 1998). These are naturally occurring pathogens that are specific for a single or a few related insect species making them environmentally safe (produce no toxic residues, harmless to non-target organisms such as beneficial insects and vertebrates, pests do not exhibit major resistance). These bioinsecticides can provide short term to long term control and are quite effective even at low dosages and the insecticidal action is only accomplished when the virions actually infect their respective target/specific host. Baculoviruses, belonging to family baculoviridae, are large enveloped viruses exclusively pathogenic to arthropods including lepidopteran, dipteran, hymenopteran insects and decapoda crustaceans with a large, double stranded covalently closed circular DNA (varying from 80 to 180 kbs), which is enclosed within an occlusion body composed of polyhedron (Vail *et al*, 1999).

Baculoviruses

- target specific viruses infecting and destroying a number of important plant pests (lepidopterous pests of cotton, rice and vegetables)
- kill pest by lysis during increased infection
- limited use in small areas due to difficulties in large-scale production
- commercially not available in India
- produced on a small scale by various IPM centers and state agricultural departments.

1. Nuclear Polyhedrosis Viruses (NPVs): Virions are protected in the environment by proteinaceous occlusion bodies, produce polyhedra containing single or multiple enveloped nucleocapsids. These infect lepidopteran larvae (caterpillars), primarily *Autographa californica* NPV (AcNPV), which have been studied intensively at the molecular level.
2. Granulosis Viruses (GVs) produce capsule shaped polyhedra called granules occulding single virion.
3. Non Occluded Viruses (NOVs) have one nucleocapsid per envelop and are not occluded.

Genetically engineered baculoviruses: Since baculoviruses require long duration (more than a week) during infection and further killing of the insect larvae, this duration has to be shortened as larvae remain quite active to voraciously devour upon the plant parts. The manipulation of the baculovirus genomes to overrule this negative would enhance the speed of kill due to combined action of baculoviral pathogenicity with the insecticidal action of toxin, hormone or enzyme. Recombinant baculovirus technology has been used to improve the insecticidal qualities of baculoviruses and employed for the expression of foreign proteins. In general the foreign genes are inserted into baculoviral genome into the polyhedron gene (*polh*) locus that encodes the occlusion body protein however other gene insertion loci (*p10*) or the nonessential region of the baculoviral genome have been identified since field stability of the recombinant virus requires intact *polh* loci. Genetic engineering could be performed using genes of various origin as Bt gene, scorpion toxin (BelT/AaIT) gene, straw itch mite toxin (TxP1) gene, insect hormone genes as diuretic hormone from *Manduca sexta* that alters the larval fluid metabolism, eclosion hormone associated with ecdysis from *Manduca sexta* causing initiation of the eclosion behavior in the inoculated larvae, ecdysteroid UDP-glucosyltransferase (egt), juvenile hormone esterase (JHE) expression in baculovirus results in stopping of larval feeding and commencing of moulting stage. A more detailed version of the above recombinant baculoviral forms and their genome manipulation using various genes could be read in nice review by Bonning and Hammock (1992).

Disadvantages

- Mode of action slow taking several days to kill insect pest and even infected larvae continue to feed causing considerable damage to crop plants
- Limited host range bars use as broad spectrum insecticide
- Small commercial potential
- Production labour intensive and not easily amenable to large scale commercial automation
- Virulence of baculoviruses depends on age of the larvae (infection sensitivity decreases with increasing age of larvae)

Field efficiency and application of Baculoviruses

Baculovirus formulations are of several types that may include the liquid forms that could be easily sprayed on the standing crops or as granular/powdery (wettable) dust types. The most commonly utilized methods to apply these formulations include the air sprays, boom-type sprayers or high pressure equipment. Under field conditions baculoviral application results in infection of the larval forms feeding on the plant parts and then the virus reaches the target site i.e midgut of the insect larva, replicates in the gut cells and reach the haemocoel from where these virions travel to infect new cells upon release of progeny virions by budding. It takes 7-8 days to approximately two weeks till the infection spreads and larva dies. Forte *et al* (1999) studied the pathogenicity of a granuloviruses towards Spruce budworm *Choristoneura fumiferana* and observed decrease in the spruce budworm larvae in the 4th instar stage of their development. Alternate sprays of baculoviral formulations and *Bacillus thuringiensis* could decrease the larval infestation better than the single use of the two types of biopesticides (Jani *et al,* 2001). Escribano *et al* (1999, 2000, 2001) evaluated the role and effectiveness of inoculation of Sf MNPV isolates on larval forms of *Spodoptera frugiperda.* A similar study was undertaken by Ebling *et al* (1997) regarding pathogenicity of NPV to forest tent caterpillar.

Two major disadvantages in their use are instability towards UV radiations or sensitivity to exposure to sunlight and slow rate of mortality of the insect pest after inoculation. Several approaches could be used to overcome/address these disadvantages attached to popular use of baculoviruses i.e. use of UV protectants to guard baculoviruses from UV ray inactivation, addition of adjuvants in baculoviral formulations to increase the site of action and proper adherence to the plant surface that may include the use of surface active agents like spreaders, stickers, emulsifiers, buffers, gustatory stimulants etc (Bonning and Hammock, 1996). However the baculoviral rate of larval mortality could be enhanced by use of genetic engineering techniques that would manipulate the baculoviral genome as expression vectors expressing novel or conjugated proteins that would increase the per cent effectiveness of viral agents as bioinsecticide using recombinant DNA technology.

c) Entomopathogenic Fungi : Inspite of the recent advances in insect pathology, the study of mycoses caused by entomogenous fungi has held a relatively modest position. Association of fungi with insects is well known and over 750 species of nearly 100 genera are found in division Eumycota and in the following subdivisions: Mastigomycotina, Deuteromycotina, Zygomycotina, Ascomycotina and Basidiomycotina (Tanada and Kaya, 1993). Among these fungal classes many entomopathogenic fungi, especially those in the order Entomophthorales, are responsible for epizootics that often successfully regulate pest insect populations and the most common method of employing fungi for insect control is through inundative means (Marrone, 2002). Some of the genera that have been most intensively investigated for development of mycoinsecticides are *Beauveria, Metarhizium, Verticillium, Poecoelomyces, Nomuraea, Erynia* etc. The first two genera i.e. *Beauveria* and *Metarhizium* have been identified from 700 and 300 species of insects respectively and have been used on a large scale over a number of years in different countries (Narayanan, 2002).

General mode of action of Mycoinsecticides

Differing from bacteria and viruses, fungi can infect insects not only through the gut but also through spiracles and particularly through the surface of the integument. This property leads directly to the theoretical possibility of infecting insects independently of their feeding activity (Ferron, 1978).

With most entomopathogenic fungi, disease development involves:

- Attachment of the infective unit e.g. conidia or zoospores to the insect epicuticle.

- Germination of infection unit on the cuticle.
- Penetration of cuticle, either directly by germ tubes or by infection pegs from appresoria.
- Multiplication of yeast phase hyphal bodies in the haemocoel.
- Production of toxic metabolites.
- Death of the host
- Growth in mycelial phase with invasion of virtually all host organs.
- Penetration of hyphae from the interior through the cuticle to the exterior of the insect.
- Production of Infective units on exterior of the insects.

Omission/failure of these fungi at first four steps will produce low virulent fungi though they might have high toxin biosynthetic capability (Arora and Dhaliwal, 2001). *Beauveria bassiana* is a common soilborne fungus that occurs worldwide and attacks a wide range of both immature and adult insects (Khachatourians *et al*, 2002). It causes white muscardine disease and is useful in control of japanese corn borer, white flies, aphids, grasshoppers, colarado potato beetle, japanese beetle, boll weevil and codling moth. Sheeba *et al* (2001) have reported the efficiacy of *Beauveria bassiana* for control of rice weevil. Similarly another epizootic fungi used as bioinsecticide is *Metarhizium anisopliae* which is a soilborne fungus that causes green muscardine disease and could be used to control locusts, grasshoppers, termites, curculionids and scarbeids. Fungal activity is positively associated with high humidity and rainfall with infection levels that remain higher under well watered conditions for the introduced resting spores (Kaaya and Hassan, 2000). Lacey *et al* (2001) have reported the aspect of inoculation of mycoinsecticide *Entomophaga maimaiga* on all larval instars of gypsy moth that were susceptible to infection by *Entomophaga maimaiga*. The techniques regarding the mass production of these entomopathogenic fungi are not upto the large scale of production standards as most species of entomophthoralean fungi are relatively difficult to produce and their primary conidia are short lived, making timing of inundative applications difficult or impossible. Development of effective methods for production of resting spores and competent mycelia of entomophthoralean species will ultimately increase the utility of these fungi (Lacey *et al*, 2001).

d) Other bioinsecticides: Apart from the most popluar Bt, baculoviral and fungal bioinsecticides other types are also being used for certain special types of insect pests. The general classes included are the protozoans, nematodes and even insects parasitizing on the pest insects as well as some insect chemicals against the biochemical processes involved in insect signaling/ communication and reproductivity (biochemical insecticides).

i) Protozoa: Insects get infested and get killed by many protozoan diseases caused by chronic infection of host specific protozoan genera and thus these eukaryotic microbes comprise an important regulatory role in insect population. The general mode of action of these entomopathogenic protozoans is through debilitation of the overall fitness and reproduction and thereby decrease in host vigor and longetivity by acting as chronic debilitating agents. *Nosemia fumiferarae* is one such typical protozoan being exploited for the management of spruce budworm. 'Noloc' is the formulation based on *N.locustae* produced by a private company M/s Sandoz Inc., is used against grasshoppers. *Varimorpha necatrix*, is another protozoan that infects 36 lepidopteran pests and has peculiar abilities of high virulence and wide host range making it an attractive candidate for functioning as better microbial control agent.

ii) Nematodes: Nematodes parasitizing on insects either alone or sometimes in combination with bacteria may be used as bioinsecticides. A plethora of nematodes belonging to more than 30 families are observed to be associated with insects and other vertebrates among which members belonging to 7 nematodal families have been most studied and are observed to have active bioinsecticidal activities. The nematodes share a symbiotic relationship with the prokaryotic bacterial genera which equip these bacteria harboring nematodes to be highly virulent in behavior so as to kill its host within 48 h by the action of the mutualistic bacteria. Bacteria involved in this interaction belong generally to genus *Xenorhabdus*, with *X. luminescens* and *X. nematophilus*. Bacteria alone cannot do any damage to the insect host, while nematodes in the absence of bacteria reproduce very poorly, and therefore, fail to cause any pathogenicity. It is the mutualistic nematode bacteria complex that enhances nematode reproduction and ultimately kills the insects. (Kaya and Gaugler, 1993). *Neoplectana carpocapsae* vectors a bacterium that builds up in the body cavity of the insect and can cause septicemia followed by insect dead. The general insects pests that are targeted by nematodal biopesticides include the soil insects and those that live under cryptic habitat conditions as citrus root weevil, black cutworm etc.

These entomopathogenic nematodes could be reared and mass produced on artificial chemically known solid/liquid monoxenic media or by *in vivo* cultivation using suitable host insect. However the generally field applied/commercial formulations that demand longer shelf life of the product at room temperature (at least 6 months) and ease of application over several acreage of land exist in form of wettable dispersible granules containing infective juveniles. The five species of nematodes that are being sold commercially as bioinsecticide are *Steinernema carpocapsae, S. riobravis, S. feltiae, Heterorhabditis bacteriophora and H. megidis.*

The nematodes being sensitive to several biotic and abiotic factors could be genetically manipulated for enhanced killing by virtue of better resistance to altered environmental conditions particularly high temperature stress, increased virulence and search capacity.

Trichogramma- insect against insect

- egg parasitic minute wasps
- controls lepidopteran pests (vegetable/fruit pests, sugarcane internode borer, pink bollworm and sooted bollworms in cotton, stem borers in rice) by devouring of eggs of pests by hatched larvae of *Trichogamma*
- most popular biocontrol agent in India
- globally produced at large scale (>10 biological factories produce about 50 billion *Trichogramma* and other parasites per season in the former Soviet Union, >50 commercial insectaries in USA/Canada, several communes in China) at a large scale.
- No large scale production in India
- small scale production for local use, mostly by sugar mills and cooperatives, state agricultural departments, IPM centers, agricultural universities and some companies through direct selling at a small scale.
- *Trichogramma* eggs have to be used within a short period (before the eggs hatch). This limits their production and marketing on a large scale, and is also the reason why *Trichogramma* is not sold through dealers and shopkeepers

iii) Biochemical insecticides: These are naturally occurring substances that control pests by nontoxic mechanisms and in contrast to conventional pesticides that directly kill or inactivate the pest. Biochemical pesticides include substances, such as insect sex pheromones, that interfere with mating (mating disruption), as well as various scented plant extracts that attract insect pests to traps. If an area is saturated with the pheromone of the pest insect, the message given out by individual insects is swamped that vigorously affect the mating process which now entirely depends on chance juxtaposition of a male and a female type of the target insect. Thus either no progeny i.e. eggs are laid by the unmated insects or even if the eggs are laid they are infertile.

BIOHERBICIDES

Chemical herbicides are the synthetic compounds that are applied to the standing crop to get rid of the unwanted competitor weed plants growing along with the crop plants. The weed plants exhibit more efficient absorption of nutrients and thus compete for nutrition and resources with the crop plants leading to suppression of the growth of the plants and hence decrease in crop yield. Thus herbicides are prerequisite for weed control and increased crop production. Since chemical herbicides are not target specific they pose toxic effects to the crop plants also. This calls for use/application of herbicides of biological origin i.e. the bioherbicides.

The bioherbicide approach has promised effective weed management in cropping systems where the classical approach (using exotic natural enemies) is largely unsuitable. The overriding principle of the bioherbicide approach has been that a host-specific, coevolved natural enemy can be used as a bioherbicide when applied in simple formulations at inundative levels (Hallet, 2005).

Bioherbicides are biocontrol agents applied for the control and eradication of weed plants and the basic purpose as well as mechanism of control is similar to the conventional chemical herbicides (Bishop, 2000). The bioherbicide tactic differs from the classical biocontrol method of weed control in which the dissemination and reproduction of the biocontrol agent is left largely up to nature. Several types of bioherbicides have been proposed and utilized for biological control of weeds such as natural plant products i.e. allelochemicals and mycoherbicides (using active fungal cultures for target specific control). Mycoherbicides have gained impetus regarding their application and production due to the target host specificity of the plant pathogenic fungi as well as the culturability of these microbes under standard lab conditions without being fastidious in nutritional requirements for their growth and mass production. Many attempts to produce bioherbicides for pastures and dryland agriculture have been thwarted by the pathogen's requirement for extended periods of dew or rain for efficacy. In addition, mass production of filamentous fungi may pose other problems. Researchers are trying to obtain more lethal pathogens without altering the host-specific nature of the invading fungal species. The most popular mycoherbicide would be the one that possesses excellence in terms of specificity and faster biocontrol of weeds and the most sort-after candidate is plant pathogenic fungi *Fusarium oxysporum* Sclecht., (Thomas *et al,* 1998). This fungus has more than ninety known forma speciales that are plant host specific to the tune of single genera or even single species (Brayford 1993). The basic mechanism of mycoherbicidal control is invasion of vascular tissue once host tissues are encountered resulting in stunting or killing of the invaded plant, however the mortality is less (around 25%) under field conditions. Thus to increase the mortality rate of the weed plants it is prerequisite to increase the virulence of the fungus. Tiourebaev *et al* (2000) have reported improvement in virulence of mycoherbicides by method of amino acid excretion and suggested incremental increase in virulence by selection for additional excretion of the same or different amino acid. Green (2003) has reviewed the potential of use of fungal herbicides for biocontrol of forest weeds that compete with the young tree saplings in woodlands as well as decrease growth in commercial tree plantations in United Kingdom and reported the greatest potential for application of a wood-rotting fungus as a bioherbicide stump treatment for *Rhododendron*. Ortiz-Ribbing and Williams (2006) have reported the use of two fungal pathogens for biological weed management of several *Amaranthus* species, some of the biotypes of which have developed resistance to multiple herbicide families. They observed 80 to 100% seedling mortality for *A. albus* and *A. blitoides,* 14 to 15 days after transplantation for the mixture or *M. amaranthi* alone, in greenhouse and field trials.

Mass production of mycoherbicides is performed either as submerged fermentation or by growing on solid substrate as wheat, oat grains, straw etc (Auld and McRae, 1997). In the bioherbicide tactic, massive doses of inoculum are applied directly to all the target weeds and thus it requires large numbers of viable propagules and a satisfactory method of storing them so that their viability is maintained until they are required for application. Two commercial bioherbicide products, Devine and Collego came onto the market in USA in the early 1980s. Since then, although new bioherbicides may have been patented and some have been registered, none have become commercial products. Difficulties in mass production of propagules have often occurred and in the vast majority of cases a practical formulation to overcome the dew requirement of fungi has not been found.

This facet of IPM strategies requires more research that should be conducted with virulent, broad host range organisms, and more effort should be devoted to develope techniques for the cultural and genetic enhancement of bioherbicidal organisms (Hallet, 2005).

BIOFUNGICIDES

Incidences of plant diseases has been controlled by agronomic practices that include crop rotation and use of agrochemicals and by breeding new strains and varieties that contain new resistance conferring genes. Among the pests attacking crop plants fungal pathogens contribute towards considerable losses per annum and thus a large amount of fungicides are sprayed to combat such infections thereby increasing the cost of production and financial burden to the farmers.

Biofungicides are the type of biopesticides used to curb and control fungal plant pathogens and include introduction/ inoculation of microbial live cells as well as engineering of the genes resulting in production of compounds that lead to fungal pathogen control or generation of a hypersensitive response through resistance genes/manipulation of genes involved in systemic acquired resistance. The genetic engineering facet would be discussed under section transgenics for fungal resistance. The general inoculant used and supplied commercially as biofungicide is *Trichoderma* sp however newer candidates for being better biofungicides are now searched, reviewed and analyzed to check efficacy upon field release.

Lahdenpera (2003) reported the use of *Streptomyces griseovirdis* K61 (tradename Mycostop) powdered formulation of dried spores and mycelia of the soil actinomycete; as potent biofungicide having ability to curb damping off of cauliflower caused by *Alternaria*. Apart from this fungal disease Mycostop was observed to be effective against other seed pathogens as *Phoma, Fusarium, Helminthosporium, Botrytis, Penicillium* etc. *Trichoderma harzianum* (Rootshield and TopShield) and *Trichoderma virens* GL-21 (SoilGard 12G) are formulations of a selected strain of these common soil saprophytic fungi. *Ampelomyces quisqualis* M-10 (AQ10 Biofungicide) is used as water-dispersible granule formulation helpful to control *Peronospora* genera of fungi involved in causing powdery mildews. The biocontrol or antifungal activity of the inoculated fungi for other target pathogenic fungi could be monitored by using confocal laser scanning microscopy (Cornick *et al*, 1999)

Trichoderma

- active against soil born root rot fungal pathogens
- useful for dryland crops (groundnut, black gram, green gram, chickpea)
- marketed by three companies in India

Plant-incorporated protectants (PIPs)/Transgenic crops

In this present era of big-bang in information, protocols, procedures and knowledge on molecular biology and genetic engineering there are increasing incidences of development of engineered crop varieties having desired genes for particular function. In 2000, four countries grew 99% of the global transgenic crop area including USA, Argentina, Canada and China. Argentina ranked second with 10.0 million hectares (23% of the global area), which represents an increase of 49% compared to the figures of 1999. The other major players were the USA (68%), Canada (7%) and China (1%). The remaining 1% is shared by a variety of developed and developing countries, including Uruguay and Mexico (James, 2000).

The last two decades have witnessed increased amount of information regarding the genomic size, type and molecular characteristics of organisms spanning over range of classes and families. Moreover enormous progress has been made over the past decade regarding the cellular level cross talk and highly complex molecular events during plant-microbial/pathogenic interactions. The vivid molecular based genomic/proteomic analysis of the plant and plant associated/pathogenic microbes have enabled molecular biologist to manipulate plant genome to render the progeny plant usher new products providing resistance to an array of pathogens. The genetic transformation of the crop plant using foreign genes from organisms of biopesticidal importance requires various basic techniques and tools *viz* target and candidate genomes, vector to carry gene of interest from nurturing organisms to host genome, modification of the target gene enabling it to accomplish expression under eukaryotic cellular conditions, method to deliver mosaic plasmid in target cell and protocols to identify transformed cells.

i) Transgenics for insect resistance: Many insecticidal proteins and molecules are available in nature, which are effective against agriculturally important pests. Research regarding the use of known/novel insecticidal genes from varied classes and families spanning over prokaryotes to eukaryotes have been searched and tried on. Corzo *et al,* (2000) have reported the isolation and synthesis of four novel insecticidal α-paulotoxin IT from the spider (*Paracoelotes luctuosus* (Amaurobiidae)) venom and exhibited significant toxicity against lepidopteran crop pest *Spodoptera litura* with LD_{50} values in the 9 to 50 mg/g of insect range. However the most talked about and applied gene is *B. thuringiensis* (Bt) for controlling *Helicoverpa armigera,* apart from use of other gene(s)/gene products as *Photorhabdus luminescens* insecticidal proteins, insecticidal viruses, plant proteinase inhibitors, lectins, chitinases or other metabolic enzymes. The major pests resulting in pre and postharvest crop damage include the lepdiopteran, orthopteran or hemipteran members. A detailed work on use of transgenic resistance to insects in crop improvement has been published by Sharma *et al* (2000) and is available online.

Since its cloning in 1981, Bt gene has travelled a long way contributing a lot towards creation of transgenic plants having Bt gene. Plants belonging to diverse genera and families as rice, broccoli, apples, soybean, maize, cotton, tobacco etc were modified to nurture and express this prokaryotic gene so as to curb various respective insect pests. The Bt gene codes for the insect gut toxin and follows the same pathway as that being followed by the ingestion of Bt spores or Cry protein.

Mode of action

- cry proteins (derived form N-terminal half of protoxin) get solubilized in midgut at high pH, releasing δ-endotoxins
- N-terminal toxin domain (amino acids 1 to 279) region is composed of alpha-helices, which are considered important in penetrating the peritrophic membrane.
- C-terminal region (amino acids 461-695) and highly variable region (amino acids 280-460) are considered important in toxin specificity by coding for open beta-sheets that bind to glycoprotein receptors in the midgut.
- gut proteases solubilize inactive crystalline protoxins by cleaving nearly 500 amino acids from the C-terminus of 130 kDa protoxin and 28 amino acids from the N-terminus, leaving 55 to 65 kDa protease resistant active core comprising the N-terminal half of the protoxin
- active toxins bind to the specific receptors (of a wide range glycoproteins, membrane proteins, aminopeptidase) located on the apical brush border membrane of the columnar cells.
- toxin inserts irreversibly into the plasma membrane of the cell after binding with receptor leading to lesion formation.
- alpha-helices can penetrate the apical membrane to form an ion channel allowing rapid fluxes of ions (K^+/cations/anions/neutral solutes as sucrose/water) i.e. Bt toxins lead to cessation of K^+ pump that leads to swelling of columnar cells and osmotic lysis.
- Disruption of gut integrity leading to insect death due to starvation or septicaemia (occurs after larval death due to toxins).

This very idea of Bt gene expression in target plant to control insect pests seems to be straight forward approach but technically it requires overruling of several hindrances to perform the action as an efficient bioinsectical product. The prokaryotic gene has to be modified from AT-rich to GC-rich bases to increase toxin expression (10 to 100 fold), addition of eukaryotic recognizable promoter (constitutive CaMV35S/ ubiquitin) and terminator gene that may/may not exhibit tissue specific expression and delivery of reconstructed gene on a plasmid or other vector by *Agrobacterium* or biolistic methods.

The major problem is that of the expression that never remains optimal to exhibit above average to phenomenal mortality rate of the voraciously feeding insect larvae. This could be overruled by using truncated gene, different promoter/ enhancer sequences and fusion protein.

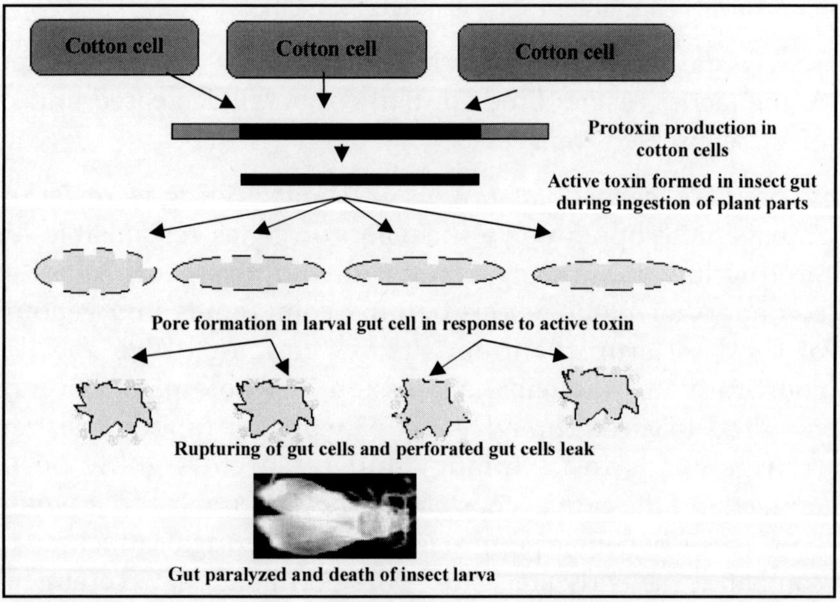

Mechanism of action of transgenic Bt cotton for the biocontrol of cotton bollworm *(For a coloured version of this figure, see plate section, page 306)*

Construction Bt transgenics

- ❒ Gene converted from AT-rich (typical of bacteria) to GC-rich (typical of higher plants) to increase toxin expression altered from the natural gene to make the gene more active in the crop and in many instances the amino acid sequence of the toxin is altered to make the toxin more soluble in the plant cell.

- each toxin is normally modified in its DNA sequence from the natural toxin by the introduction of regulatory sequences such as introns, polyA signals, promoters and enhancers.
- expression of Bt gene in higher plants requires a recognizable promoter and a terminator sequence bracketing the Bt gene.
- popular constitutive promoters include cauliflower mosaic virus (CaMV35S) and ubiquitin. Tissue specific promoters include PEPC (phosphoenolpyruvate carboxylase) (for green tissue) and maize pollen specific promoter
- vector size ranges from 5000 to 11000 bp depending on Bt gene and the promoter incorporated into the vector
- vector delivered into nucleus achieved by using *Agrobacterium* mediated transformation and biolistic methods

Tissue specific expression could be useful and enunciated as the tactics to avoid the dilution or overall decreased impact of the expressed transgene.

ii) Transgenics for fungus resistance: Fungus resistant transgenic crops involve inclusion of genes responsible for production of antifungal molecules in any genre/class of secondary or pathogenesis curbing compounds into genome of the host crop plant and its systemic or tissue specific constitutive or inducible expression in whole plant body or required tissues. Though most of work on fungus resistant transgenics is in its infancy and research is going on to understand the prime steps of host-pathogen relation it would be a more beneficial approach to avoid fungal attack or to combat at the early stages of spore germination. Several gene products involved in systemic or induced defense response have been reported to infer fungus resistance *in vitro* and could be utilized for creation of a better transgenic such as phytoalexins, pathogenesis-related proteins (five types of PR proteins), ribosome inactivating proteins (RIPs) of plant origin, small cystein-rich proteins, lipid transfer proteins, non-plant antifungal proteins, polygalacturonase inhibitor proteins etc. However another approach for creating plants capable of resisting pests is to develop transgenics having genes engineered for hypersensitive response. Certain genes (resistance genes) that encode for antifungal compounds as PR proteins (chitinase enzyme), phytoalexins, toxins and other similar compounds could be engineered in candidate plant genome and such transgenic plants would express that particular gene(s) thus imparting resistance to that fungal pathogen (Grover and Gowthaman, 2003).

Future Perspectives

The future perspectives in biopesticides are leaning towards development and release of transgenic crop plants that themselves have the acquired ability to annihilate the pest attack. Several molecular tools/techniques and protocols are helping in pinpointing a relationship between the introduced foreign gene (of biopesticidal action) and the candidate host line/variety that exhibits maximum expression of the introduced gene and thus enhanced mortality of the target pest. The correlation studies between the native resistance genes in plant and the introduced transgene say Bt gene have showed that resistant variety with transgenic insertion exhibits higher mortality than susceptible line having transgene (Walker *et al,* 2004).

Thus the strategy of gene pyramiding using either multiple Bt genes, Bt and unrelated transgenes, or Bt and native plant genes could help rescue the development of resistance to known arsenal of biopesticides. Expression of multiple Bt genes should increase mortality of insects that have evolved resistance to one of the Bt toxins, and cotton (*Gossypium hirsutum* L.) cultivars carrying cry2Ab and cry1Ac have been developed (Adamczyk *et al,* 2001). Walker *et al* (2004) have used a multiple resistance gene pyramiding strategy to obtain the soybean variety possessing two quantitative trait loci (QTLs) exhibiting positive correlation with the Bt transgene.

Advantages of biopesticides

- eco-safe, inherently less harmful than chemical pesticides
- environmental compatibility as cause no contamination of ground and surface water aquifers
- more target specific affecting only the target pests and their close relatives
- conservation of natural predators/parasites of pests
- unharmed beneficial pollinator insects (honey bees)
- do not cause resurgence of older pests of crops
- no pest resistance developed
- genetically modifiable
- often effective in small quantities cutting down the cost of pesticide application and financial burden to farmers
- chemical compatibility as can be used with some chemical insecticides
- decompose quickly and do not leave problematic residues (do not biomagnify)
- reduced number of applications
- increased yield and quality improvement with added benefits for being organic in origin
- higher acceptability of agribased products particularly fruits, vegetables, grains and tea in global market
- ease of application as can be applied with standard spray equipment and do not require the purchase of highly specialized and dedicated sprayers and not toxic during application and handling thus beneficial for ignorant and unskillful rural Indian farmers (Hoffmann and Frodshan, 1993)

Based on defoliation in the field and larval weight gain on detached leaves, lines carrying a combination of cry1Ac and the PI 229358 allele at a QTL on linkage group M were significantly more resistant to the lepidopteran pests, including the Bt resistant TBW strain, than were the other lines.

Effect of transgenic crops on soil microbiota: Transgenic plants carry genes and produce compounds foreign to their environment that concerns about their environmental use and potential ecological effects. The issue of direct and indirect effects on nontarget organisms and ecosystems is particularly important because many transgenic plants are being developed that have new or enhanced antimicrobial properties for protection from phytopathogens. These transgenic plants express such antimicrobial compounds as chitinases, glucanases, lysozymes, thionins, defensins, and systemic acquired resistance (SAR) gene products, harbor antibiotic, herbicide/weedicide resistance genes and produce novel toxins for pest resistance. Microcosm and field studies showed that exposure to transgenic plants produced changes in the population levels and composition of some soil and plant microorganisms (Dunnfield and Germida, 2004). Microcosm studies also revealed *B.t.k.* endotoxin persistence at low levels to be of ecological concern as repeated use of *B.t.k.* endotoxin-producing plants lead to accumulation of the endotoxin and would potentially adversely affect soil organisms. GM potatoes consistently altered the physiological profile of the rhizosphere microbial community at harvest, but the effect did not persist from one season to the next (O'Callaghan and Glare, 2001). The transgenic crops alter both the population levels and species composition of bacteria and fungi. Moreover transgenic plants cause transient but significant increase in levels of culturable, aerobic bacteria and fungi. Transient changes may be manitested as stimulated bacterial and fungal populations (Donegan *et al.,* 1995) or decreased diversity. There could be a higher incidence of the plant pathogen *Verticillium dahliae* on the transgenic *B.t.t.*-producing plants owing to their enhanced longitivity. Transgenic plants that produce opines can influence the composition of rhizosphere bacteria by altering the populations of rhizosphere bacteria due to selective utilization potential of opines by them. A delay in the onset of colonization, and a decrease in the level of colonization, by the arbuscular mycorrhizae *Glomus mosseae,* was observed on transgenic tobacco plants with chitinase expression. Light microscopy revealed distinct differences in the fungal structures on the transgenic plants as compared to the control plants. Pesticidal proteins produced in transgenic plants can persist in soil and binding of these proteins to soil particles can protect them from biotic degradation. Even the DNA in transgenic plants can persist in a field environment for several months. There is potential for horizontal gene transfer of antibiotic resistance from transgenic plants to microorganisms.

Conclusions

Biopesticides are the integral components of Integrated Pest Management (IPM) strategies and are efficient agencies to get rid of notorious crop pests resulting in decreasing the net profits due to damage at the pre as well as postharvest stages of the crop production. Crop pests include a wide range of insect, fungal, bacterial, viral agents, rodents, birds and other animals that tremendously damage the crop at various stages of maturation and storage among which insect pests are most noteworthy. Entomopathogens are the potential candidates for the management of insects pests and the need of the hour is to develop bioinsecticides having high lethal effect at low concentration and in less time, besides being cost effective. The use of bioinsecticides will not only help in preventing the dumping of thousands of tons of agro chemical on the earth but alsowill provide the pesticide residue free food and a safe environment to live. Research and development of protocols and processes in formulation of one such bioinsecticide includes the combination of two or more effective lethal processes and tailoring these into one organism so that the probability of development of resistant strains lessens to negligible due to little morbidity and effective mortality factors. Several government and public agencies are manufacturing biopesticides at the commercial scale and are providing farmers these weapons for fighting against crop pathogens and pests. Genetically modified microbes and transgenic crops are the recent episodes in the chapter of IPM strategies and are gaining popularity regarding their efficacy in eradicating pests but stern concerns are evident regarding the potential environmental hazards (vertical/horizontal transfer of prokaryotic genes) of their long term use.

References

Adamczyk JJ, Adams LC, Hardee DD (2001) Field efficacy and seasonal expression profiles for terminal leaves of single and double *Bacillus thuringiensis* toxin cotton genotypes. *J Econ Entomol,* 94:1589-1593.

Arora R and Dhaliwal GS (2001) *Integrated Pest Management. Concept and Approaches.* Microbial Control, Kalyani Publishers, pp.197-233.

Auld BA and McRae C (1997) Emerging technologies in plant protection-Bioherbicides. *In: Proc. 50th N.Z. Plant Protection Conf. 1997, 191-194*

Barewale RB, Gadwal VR, Zehr U and Zehr B (2004) Prospects for Bt cotton technology in India. *AgBioForum,* 7(1&2):23-26.

Bishop A (2000) Bioherbicides: New products and new modes of action. Tas- Regions 6(3): 41-42.

Bonning B C and Hammock B D (1996) Development of recombinant baculoviruses for insect pest control. *Annu. Rev. Entomol.* 41 : 191-210.

Bonning BC and Hammock BD (1992) Development and potential of genetically engineered viral insecticides. *Biotechnol. Genetic Engineering Revs,* 10:455-486.

Brayford D (1993) The Identification of *Fusarium* species. Workshop Manual. International Mycological Institute, Bakeham Lane, Egham, Surrey, UK 18-22 October, 1993. CAB International.

Commandeur P and Komen J (1992) Biopesticides–options for biological pest control increases. *Biotech and Develop. Monitor,* 13 : 6.

Connick Jr. WJ, Ingber BF, Goynes Jr WR, Daigle DJ, Jackson MA and Lewis JA (1999) Microscopy as a research tool in the biocontrol of agricultural pests with fungal pathogens. *American Arboratory,* (Nov):13-16.

Corzo G, Escoubasi P, Stamkiewicz M, Pelhatez M, Kristensen CP and Nakajima T (2000) Isolation, synthesis and pharmacological characterization of ä-palutoxins IT, novel insecticidal toxins from the spider *Paracoelotes luctuosus* (Amaurobiidae). *Eur. J. Biochem.,* 267:5783-5795.

Dhaliwal GS and Pathak MD (1993) Pesticides in the developing world: a boon in ban. In: Dhaliwal, GS. and Singh, B. (eds). *Pesticides:*

Their ecological impact in developing countries. Commonwealth Publishers, New Delhi.

Donegan KK, Palm CJ, Fieland VJ, Porteous LA, Ganio LM, Schaller DL, Bucao LQ and Seidler RJ (1995) Changes in levels, species, and DNA fingerprints of soil microorganisms associated with cotton expressing the *Bacillus thuringiensis* var. *kurstaki* endotoxin. *Appl Soil Ecol,* 2:111-124.

Dunfield KE and Germida JJ (2004) Impact of genetically modified crops on soil- and plant-associated microbial communities. *J Environ Qual,* 33:806-815.

Ebling PM and Kaupp WJ (1997) Pathogenicity of a nuclear polyhedrosis virus to forest tent caterpillar, *Malcosoma disstria. The Canadian Entomologist,* 129: 195-96.

Escribano A William S T Goulson D Cave R D Chapwan J W and Caballero P (1999) Selection of nucleopolyhedro virus for control of *Spodoptera frugiperda* : Structural, Genetic and Biological comparison of four isolates form the Americas, *J Eco Entomol,* 92(5):1079-1085.

Escribano A, Williams T Goulson D, Cave RD, Chapman JW and Caballero P (2001) Consequences of interspecfic competition on the virulence and genetic composition of a nucleopolyhedrovirus in *Spodoptera frugiperda* larvae parasitized by *Chelonus insularis. Biocontrol Science Technology,* 11:649-662.

Escribano A, Williams T, Goulson D, Cave RD and Caballero P (2000) Parasitoid-pathogen-pest interactions of *Chelonus insularis, Campoletis sonorensis,* and a nucleopolyhedrovirus in *Spodoptera frugiperda* larvae. *Biological Control,* 19:265-273.

Ferron P (1978) Biological Control of insect pests by Entomogenous fungi. *Annu Rev Entomol,* 23:409-42.

Forte AJ, Guertin C and Cabana J (1999) Pathogenicity of a granulovirus towards *Choristoneura fumiferana* (Lepidoptera:Tortricidae). *The Canadian Entomologist,* 131: 725-27.

Green S (2003) A review of the potential for the use of bioherbicides to control forest weeds in the UK. *Forestry,* 76(3):285-298.

Grover A and Gowtahman R (2003) Strategies for development of fungus-resistant transgenic plants. *Curr Sci,* 84(3):330-340.

Hallet SG (2005) What are the bioherbicides? *Weed Sci,* 53(3):404-415.

Hoffman M P and Frodshan A C (1993) Natural enemies of insect pests. Cooperative Extension, Cornell University, Ithaca, New York, 63 pp.

Iqbal M, Verkerk RHJ, Furlong MJ, Ong PC, Rahman SA and Wright DJ (1996) Evidence for resistance to *Bacillus thuringiensis* (*Bt*) subsp. *kurstaki* HD-1, *Bt* subsp. *aizawai* and Abamectin in field populations of *Plutella xylostella* from Malaysia. *Pestic Sci,* 48:89-97.

James C (2000) Global Status of Commercialized Transgenic Crops: 2000, *ISAAA Briefs,* No. 21.

Jani JJ, Godhani PH, Mehta DM and Yadav DN (2001) Developing microbial pesticides based module for control of pod borer complex in pigeon pea. *Symposuim on biocontrol based pest management for quality crop protection in the current millennium,* July 18-19, 2001, PAU, Ludhiana.

Kaaya GP and Hassan S (2000) Entomogenous fungi as promising biopesticides for tick control. *Experimental Appl Acarology,* 24:913-926.

Kaya HK and Gaugler R (1993) Entompathogenic Nematodes. *Annu Rev Entomol,* 38 : 181-206.

Khachatourians G G, Velencia E P and Miranpuri G S (2002) *Beauveria bassiana* and other entomopathogenic fungi in management of insect pests. In : Opender Kaul and G S Dhaliwal (eds.) *Microbial Biopesticides,* Taylor & Francis Inc, New York, London, PP 239-75.

Lacey L A Frutos R Kaya H K and Vails P (2001) Insect pathogens as biological control agents: Do they have a future? *Biological Control,* 21: 230-48.

Lahdenpera ML (2003) Streptomyces biofungicides in seed application. *Veradera Infoletter,* 12:1-3.

Liu YB and Tabashnik BE (1997) Experimental evidence that refuges delay insect adaptation to *Bacillus thuringiensis*. *Proc R Soc Lond B,* 264: 605-10.

Marrone PG (2002) An effective biofungicide with novel modes of action. Pesticide *Outlook,* (October):193-194.

Mishra S (1998) Baculoviruses as biopesticides. *Curr Sci,* 75 (10):1015-1025

Naranjo SE, Head G and Dively GP (2005) Field studies assessing arthropod nontarget effects in Bt transgenic crops:Introduction. *Environmental Entomolgy,* 34(5):1178-1180.

Narayan K (2002) Microbial control of insect pests. Role of Genetic Engineering and tissue culture. In: Opender Kaul and G S Dhaliwal *Microbial biopesticides.* Taylor & Francis Inc., New York, London pp. 117-180.

Neale MR (1997) Biopesticides-harmonisation of regulation requirements within EU Directive 91/414 - an industry view. *Bulletin OEPP,* 27:89-93.

O'Callaghan M and Glare TR (2001) Impacts of transgenic plants and microorganisms on soil biota. New Zealand Plant Protection 54:105-110 *cited online www.hortnet.co.nz/publications/nzpps*

Ortiz-Ribbing L and Williams MM (2006) Potential of *Phomopsis amaranthicola* and *Microsphaeropsis amaranthi,* as bioherbicides for several weedy *Amaranthus* species. *Crop Protection,* 25:39-46

Schnepf E, Crickmore N, Van Rie J, Lereclus D, Baum J, Fieitelson J, Zeigler DR and Dean DH (1998) *Bacillus thuringiensis* and its pesticidal Crystal proteins. *Microbiol Mol Biol Revs,* 62:775-800.

Sharma HC, Sharma KK, Seetharama N and Ortiz R (2000) Prospects for using transgenic resistance to insects in crop improvement. *Electronic J Biotechnol,* 3(2):76-95. cited online at http://www.ejb.org/content/vol3/issue2/full/3

Sharma MC (2003) Biopesticides: Potential for Indian agriculture. *In: Biotechnological strategies in Agro-Processing* (eds) Marwaha SS and Arora JK, Asiatech Publishers Inc., New Delhi pp. 105-112.

Sheeba G Seshadri S Raja N Janarthanan S and Ignacimuthm S (2001) Efficacy of *Beauveria bassiana* for the control of rice weevil *Sitophilus oryzae* (L) (Coleoptera:Cuculionidae). *Appl Ent Zoo,* 36(1):117-120.

Srinivasan G (1997) Panel for reduced use of chemical pesticides. Hindu Business Line, November 5, 1997.

Tabashnik BE, Dennehy TJ and Carriere Y (2005) Delayed resistance to transgenic cotton in pink bollworm. *PNAS,* 102(43) :15389-15393.

Tanada Y and Kaya H K (1993) *Insect Pathology*. Academic Press Inc., Narcourt Brace, Javanovich Publishers, San Diego

Thomas H, Sauerborn J, MhllerStver JDM, Ziegler A, Bedi JS and Kroschel J (1998) The potential of *Fusarium oxysporum* f. sp. *orthoceras* as a biological control agent for *Orobanche cumana* in sunflower. *Biol Control*, 13: 41-48.

Tiourebaev KS, Nelson S, Zidack NK, Kaleyva T, Pilge AL, Anderson TW and Sands DC (2000) Amino acid excretion enhances virulence of bioherbicides. *Proceedings of the X International Symposium on Biological Control of Weeds,* 4-14 July 1999, Montana State University, Bozeman, Montana, USA, Neal R. Spencer (ed). pp.295-299.

Vail PV, Hostetter DL and Hoffmann DF (1999) Development of the Multi-nucleocapsid nucleopolyhedro viruses (MNPVs) infectious to loopers (Lepidopetra, Noctuidae: Plusiinae) as microbial controls agents. *Int Pest Manage Rev,* 4:231-57.

Walker DR, Narvel JM, Boerma HR, All JN and Parrott WA (2004) A QLT that enhances and broadens Bt insect resistance in soybean. *Theor Appl Genet,* 109:1051-1057.

Whalon ME and McGaughey WH (1998) *Bacillus thuringiensis*: Use and resistance management. In: *Insecticides with novel modes of action: mechanism and application*, Ishaaya I and Degheele D (eds.), Springer, Berlin, pp.106-137.

Mind Bogglers

1. What are biopesticides? Give their types.
2. Why Bt products and formulations are used as most popular biopesticide products? What is the mechanism of action of Bt toxin?
3. Baculoviruses seem wonderful bioinsecticide but are hinderded due to certain reasons for wide spread use. Enlist those reasons.
4. What are the common biofungicides used and marketed? How one can monitor the activity of a fungal fungicide?

5. The future of agricultural biotechnology is transgenic crop plants. Comment on this statement.
6. What according to you would be possible problems of large scale use of transgenic crops?

10

Organic Matter Decomposition

- Introduction 258
- Associated Microflora and Fauna 260
- Mechanism of Organic Matter Decomposition 261
- Role in Sustainable Agriculture 262
- Future Perspectives 268
- References 270
- Mind Bogglers 271

Introduction

Soil forms the base of the lithosphere on earth and is a fundamental and irreplaceable versatile three dimensional dynamic system that supports majority of terrestrial plant and animal life on earth. This soil system contains mineral (sand, slit and clay forming basic soil structure) and organic matter as well as marginal air and water contents. The soil organic matter content of the soil system constitutes every fraction of matter in various stages of decomposition that once was associated with the body, body parts, exudates, tissues and cells of living organisms including plants, animals and microorganisms. It is the product of continuous cycling of the compounds synthesized by autotrophic plants among the three kingdoms *viz* microbes, plants and animals. The decomposition of the organic residues in soil is performed primarily by the resident soil microflora termed as microbial decomposers; though initially the larger and complex organic material is broken down, devoured upon, chewed, regurgitated and transformed into smaller bits by larger soil faunal forms (earthworms, beetles, springtails, and collembola, the macro (large) to mesofauna) on which soil microbial forms could act upon.

Types of organic compounds

- ❐ Complex and simple carbohydrates (starch, glycogen, cellulose, hemicellulose, pectins, gum, mucilage, cellubiose, glucose, xylose)
- ❐ Hydrocarbons as aldehydes, ketones, organic acid, pigments, phenolics (lignins, coumarins, phytoalexins, phenols, tannins, alkaloids)
- ❐ Proteins, fats, waxes, oils and resins

Thus the soil system consists of the soil formulating components and the soil textural and fertility components together integrated in several harmonious processes to help sustain life in form of plants and animals. The former components are the unique mineral elements/compounds that are added up to the existing soil system by the physical weathering processes of rocks particularly change in temperature and resulting chemical decomposition of larger fractions of rocks for formation of soil. The latter component comprises the biological activities particularly including the primal role of the plant root system, its growth, secretion of exudates, its metabolic activities as well as root associated or non-associated microbial metabolism.

Soil organic matter synthesized by a complex and dynamic system is comprised of three components

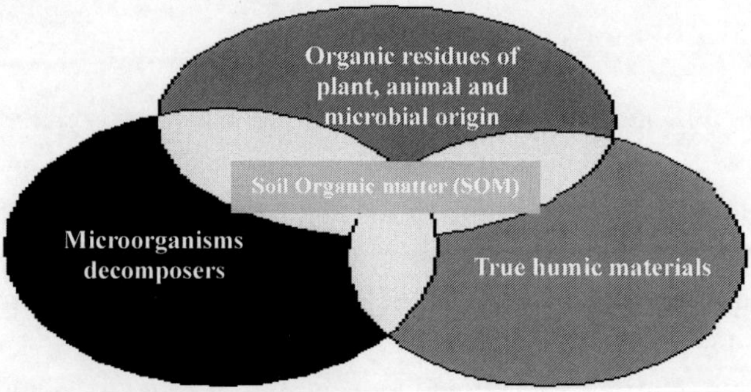

A comparison of the organic C contents of different types of mineral soils indicates that the presence of multivalent cations such as Ca (as Ca-containing minerals or exchangeable cation), Al or Fe (as amorphous Al and Fe minerals) leads to accumulation of organic C in comparison to other soil types (Sombroek *et al*, 1993). A positive relationship between soil organic C contents and high base status or presence of substantial contents of Al and Fe oxides exists, particularly the presence of reactive $CaCO_3$ in soil leads to biological stabilization of both particulate organic carbon (POC) and humus by precipitation of thin carbonate coatings (on fresh residues) and the formation of Ca-organic linkages on more decomposed humus.

The soil organic matter (SOM) accounts for the general health and productiveness of a particular soil as alterations in SOM may dramatically alter the fertility, water availability, soil compaction, susceptibility to erosion as well as several biological properties (resistance to insects and diseases) of soil of a particular region/niche. Since SOM works as a gluing agent to keep intact the soil textural elements (sand, silt and clay particles) its decrease would cause a massive change in soil compaction property (which would increase as less air passage/channels would be there in soil aggregates) or the soil structure (formation of macroagaregates due to binding action of long chain polysaccharides and other gluing compounds released by microflora during different stages of decomposition) thereby resulting in increased susceptibility to erosion/washoff by torrential rains and winds.

Why organic matter so important/what is its importance??????

- provides a carbon and energy source for soil microbes
- aids the growth of crops by improving the soil's ability to store and transmit air and water
- stores and supplies nutrients (N, P, S, K, micronutrients) required for proper plant/soil fauna growth
- retains nutrients by providing cation-exchange and anion-exchange capacities
- maintains soil in an uncompacted condition with lower bulk density that promotes water infiltration and reduces runoff
- reduces the negative environmental affects of pesticides, heavy metals and other pollutants by binding with the contaminants
- makes soil more friable and easier to work so that plant roots can penetrate the soil profile better

Associated Microflora and Fauna

The organic matter decomposition is carried on by a variety of microbes however the activity of other macroscopic organisms as nematodes, earthworms and a range of soil insects help in decreasing the higher chunks of organic material and increased formation of to smaller bits that are readily worked upon by the soil microflora (Neil and Gignoux, 2006). The dominant microflora includes mesophilic bacteria and fungi which are later replaced by thermophilic actinomycetes and fungi according to the decomposition stages. The microbial decomposer diversity varies according to the type of raw organic matter that has to be acted on (Krull *et al,* 2001) for example there occurs abundance of cellulose and lignin degrading basidiomycetous fungal genera *viz Mycena, Trametes, Stereum, Polyporus, Collybia, Citocybe, Plerotus, Marasmus* and many more if the plant organic content is in form of logs and litter rich in lignin content while the fungal genera differ for the decomposition of organic matter derived from horticultural produce that contain low concentration of lignin while more of sugar and water content. The common genera involved in decomposition of certain organic materials are being given in the following Table.

Substrate nature	**Genera of microbes**		
	Actinomycetes	**Bacteria**	**Fungi**
Cellulolytic microbes (degrade cellulose to yield glucose from macro/micromicelles of cellulose fibrils)	*Micromonospora, Nocardia, Streptomyces, Streptosporangium*	*Achromobacter, Bacillus, Cellulomonas, Clostridium, Cytophaga, Pseudomonas, Sporocytophaga, Vibrio*	*Alternaria, Aspergillus, Chateomium, Fusarium, Penicillium, Rhizopus, Rhizoctonia, Trichoderma*
Microbes attacking **hemicellulose** content including **pectin** and middle lamella constituents to release pentasugars xylose and other compounds	*Streptomyces*	*Achromobacter, Bacillus, Clostridium, Cytophaga, Lactobacillus, Pseudomonas, Vibrio*	*Alternaria, Aspergillus, Fusarium, Rhizopus, Trichothecium, Verticilluim*
Microbes decomposing aromatic compounds			
lignin	Not reported	*Flavobacterium, Pseudomonas*	*Arthrobotrys, Clavaria, Collybia, Flammula, Humicola, Lepiota, Mycena,*
Tanin	Not reported	Not reported	*Aspergillus, Penicillium*
Cutin	*Streptomyces*	*Bacillus*	*Mortierella, Penicillium, Rhodotorula*
Polysaccharide or polymer degrading microbes			
Starch (amylose+amylopectin glucose polymer)	*Micomonospora, Nocardia, Streptomyces*	*Achromobacter, Bacillus, Chromobaterium, Clostridium, Cytophaga*	*Aspergillus, Fusarium, Polyporus, Rhizopus*
Inulin (Fructose polysaccharide)	Not reported	*Cytophaga, Clostridium, Flavobacterium, Micrococcus, Pseudomonas*	*Aspergillus, Fusarium, Penicillium,*
Proteins and Nucleic acids	Not reported	*Bacillus, Clostridium, Micrococcus, Pseudomonas, Serratia*	Not reported
Complex organic acids			
Humic acid	Not reported	Not reported	*Penicillium, Polystictus*
Fulvic acid	Not reported	Not reported	*Poria*

Mechanism of Organic Matter Decomposition

The decomposition of organic matter is the general process of ecosystem and this organic matter input occurs in form of plant, animal as well as microbial biomass. The general plant residues primarily include cellulose, hemicellulose (pectin), lignin (aromatic compound encrusted in cellulose and pectin), proteins and to a small extent sugars, carbon compounds as organic acid and amino acids. In general the simpler compounds are attacked vigorously by a number of microbes while the complex aromatic compounds (polymers of substituted phenolic compounds) are attacked by few specific genera or type of microbes.

Weathering → Humus formation → Soil profile development

Natural process of organic matter decomposition

(For a coloured version of this figure, see plate section, page 306)

Role in Sustainable Agriculture

The organic farming practice is incomplete without the judicious and renewable use of the organic waste generated at death, decay or senescence of all the living organisms inhabiting earth. It is an ecological cleaning and recycling process that the organic wastes get decomposed to release the nutrients back to the soil nutrient pool. Thus it is a virtue of a natural self sustained ecosystem that decomposition occurs at its required rate to compensate for the growth and proliferation of all the organisms dependent on it. However the recent years of overexploitation resulted in disruption of the major delicate dependencies required for the self sustenance of the natural process. Now the scientists and researchers are moving towards the realization of the concept of sustainable crop production. The major facets involved in this concept related to organic matter decomposition and recycling of nutrients are :

Composting

Over the years new technological advancements and rapid urbanization resulted in creation and accumulation of urban (industrial and municipal) and agricultural waste, a major challenge regarding the management of this huge amount and variety of solid wastes. The general ways of incernation to get rid of this hiked amount of solid waste would just increase the common environmental concern of global warming and composting would be a cleaner and judicious technique for addressing such problems.

It is not advisable to use untreated raw and unstable organic wastes for addition of organic content to soil as these act as substrates for a variety of human pathogens, exhibit explicitly uncontrolled/unknown physical and chemical properties and are malodorous due to development of unwanted microflora and poor decomposition. Thus composting seems to be an important decomposition method that helps in transforming waste into wishfully useful soil amendment or soil conditioner. It could be defined as biooxidation of organic matter at an accelerated phase by action of microbes that create a thermophilic stage (45 to 65°C) by liberation of heat, CO_2 and water to finally transform heterogeneous organic material into a homogeneous and stabilized humus like product through turning or aeration.

The product of composting reactions is an environmentally safe (as abates malodors) and ecologically viable (recycles mineral and nutrition in waste products) soil amendment that make available minerals and nutrition (in form of organic matter containing C, N, P and S locked in form of plant and animal waste) to the region of its application. Composting results in conversion of complex organic forms to simpler products which could be readily utilized by other microbes as well as plants (cellulose to simple sugars, proteins to amino acids, fats and waxes to respective fatty acids) however it also results in synthesis of complex substances as humic acid and loss of labile C compounds.

The process of composting is governed by a succession of microflora and certain chemical reactions which are set in at high temperature, the microflora plays a key role during composting and these are regarded as the active agents in the preparation of nutrient medium. Various kinds of flora are encountered during the entire process of composting which could be prepared by several methods; classified according to the criteria of composting i.e. indoor or outdoor composting, long or short method of composting and organic or synthetic composting. These composting methods in anyway rely largely on the proper working, relative occurrence/dominance and subsequent succession of various types of microflora over time. Adequate knowledge of microbial succession is therefore, very important in any chosen composting method. The major rather the sole microbiological agents responsible for decomposition of carbonaceous materials are the heterotrophic microorganisms. The composition of active mycoflora of composting wastes normally shifts from predominantly mesophile (*Clasdosporium herbarium*, *Aureobasidium*, *Alternaria alternata* and *Epicoccum purpurascens*) in the early stages of thermogenesis to one of predominantly thermophiles at the peak of the heating cycle.

Successful composting depends on a number of factors such as the type of raw material being composted, its nutrient composition, moisture content, temperature, acidity or alkalinity and aeration that have both direct and indirect influence on the activities of the microorganisms as microbes require certain oxygen concentrations at high temperature and moisture content for properly decomposing the waste material (generally wheat/rice straw, cotton stems etc). There occurs a drastic decrease in the overall quality and the associated benefits of composting if these factors are not taken under consideration during composting process. This remains the basic hindrance for the use of compost formed by unimproved methods. Thus the pinpointing of composting microbial succession, inoculation of specific microflora during different stages of composting would hasten the decomposition or bidegradation potential that could imperatively yield good quality composts in shorter time spans (Taiwo and Oso, 2004).

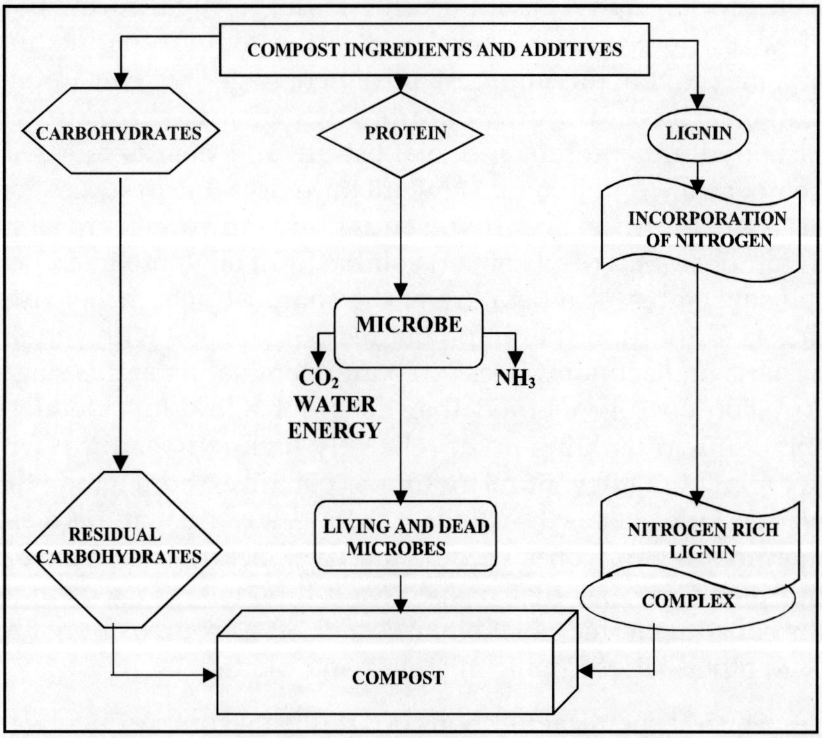

Among the various stages of composting four have been reported to occur explicitly as the routine process of microbial succession *viz* mesophilic phase (S), thermophilic phase (T), cooling phase (C) and the maturation phase (M). The thermophilic phase is important to reach as easily degradable organic matter is converted into stable matter (having humic-like substance) by passing through a thermophilic stage. Monitoring of the microbial succession is important in the effective management of the composting process as microbes play key roles in the process and the appearance of some microbes reflects the quality of maturing compost. Generally during the process of composting there occurs succession of microflora ranging from mesophilic to thermophilic organisms depending upon the level of nitrogen, temperature of incubation, stage of composting, zone of sampling etc. Mesophilic flora particularly mesophilic fungi (including primary saprophytes, opportunistic fungi, fugal forms that grow on stored materials) present in the composting mixture become active at this initial establishment stage low temperature range. These fungi degrade the straw, breakdown carbohydrates present in the substrate and heat is released. Temperature of the composting mixture rises due to succession of thermophilic organisms over mesophilic flora that decompose straw and convert soluble form of N into microbial cell substances. Temperature of the compost pile further rises due to activity of thermophilic microorganisms with certain bacteria and actinomycetes (fire fang) dominating at this stage. Any composting method that does not allow for adequate aeration and building-up of relatively high temperatures will not allow the relevant microflora to colonize and degrade the wastes. However as the compost cools down and matures these thermophilic microbes become inactive, lock up the available nutritional pool as their respective cell bodies and remain as the constituent microbial biomass and act as source of organic N to other heterotrophic microbes and plants.

Several chemical reactions occur during composting. High temperature during phase I can increase ammonia production and that ammonia increases the degradability of straw cell wall polysaccharides (Savoie *et al*, 1995). Ammonia produced due to decomposition of protein further raises the temperature of the heap and softens the straw. The soluble form of nitrogen is converted into microbial cell substances. Thermophilic microflora utilize ammonia and convert it into microbial biomass. Certain chemical changes result in the formation of N-rich lignin-humus complex as a result of which substrate becomes dark brown in colour. Maillard reactions in which sugar complex with amino acids, amines and ammonia produce carbon that are not so available to the composting organisms. Free ammonia also releases polysaccharides from the lignin matrix, making the straw carbohydrate content more available to the crop.

Vermicomposting

Vermicomposting is a biooxidation and stabilization process of organic material that does not involve thermophilic stage and is a product of heterogenous organic matter transformation reaction requiring joint action of earthworms and microorganisms where earthworms function to fragment larger chunks of organic matter to smaller forms, aerate the inhabited soil and are helpful in turning of soil. It has some added advantages over normal composting as the decomposition rates are faster than the latter, worms also help in overruling the requirement for the thermophilic phase, the end product of vermicomposting contains higher measurable nutrient levels and microbial populations, humification processes are greater and faster and above all earthworms act as natural scavengers so that they clean the end product of the probable pathogenic microbes as *Salmonella, Shigella, E.coli* and eggs of several parasitic worms and mine the vermicompost/ soil of heavy metals thereby decreasing the amount of bioavailable heavy metals.

Green manuring

Green manuring is the age-old practice of soil incorporation or tilling, spading or plowing under of forage or leguminous plants or any other green manure plants after one to one and half month of growth just after the reproductive maturation stage of development while it is still fresh for improving the physical structure and fertility of soil. The major benefits to the use of green manures in a crop rotation system include organic matter and nitrogen addition, improvement of the soil physical properties as aeration, porosity, water-holding capacity thereby facilitating penetration of rainwater, nutrient conservation, and protection of the top soil (soil surface) from erosion though it is difficult to assess the benefits of green manuring quantitatively and the degree of improvement necessary to reach a stage that helps the growing crop most or in a significant manner. Other essential macro and micronutrients such as potassium, phosphorus, magnesium and iron can also be conserved through green manuring.

Moreover microbiologically this process causes increase in the soil microbial activities particularly that of soil fungi and bacteria as well as increases soil microbial diversity and count making it microbiologically sounder and agronomically more productive. The benefits of green manuring on crop yield are most apparent during dry periods, particularly in rainfed production systems. The low-nitrogen green manures (1.5% N or less) can be effective in improving soil organic matter levels while the leguminous crops nitrogen addition is not to an extent due to rapid loss of the available N. Certain general green manuring plants include *Crotolaria juncea, Alfalfa,* black lentil, berseem, lupine, *Trifolium pratens, Melilotus officialse* (sweet clover), *Sesbania rostrata, Lathyrus sativus* etc.

Future Perspectives

Today's is the scenario of explicit development in the field of methodologies, techniques and procedures along with tools/ equipments in the study of ecological role of microbes in an ecosystem particularly the decomposers (bacteria, fungi and actinomycetes). The Microcosm studies of the decomposers revealed that these unspoken scavengers of planet earth are relentlessly diverse and have delicate intricacies as well as interdependencies to carry out the whole process of decomposition and nutrient recycling. The relationship between biotic diversity and ecosystem function has recently gained much interest among ecologists. Setala and McLean (2004) have performed a microcosm study to investigate the relation of number of fungi to system's ability to (i) degrade raw coniferous forest humus, and (ii) use resources that were either added to the systems or released into the soils after a disturbance (drought). They reported increase in diversity of the fungi resulted in increase in the CO_2 production which was unlikely to be correlated to addition of various energy sources as the latter treatment in general increased the soil respiration but does not affect the shape of the diversity-CO_2-production curve though the fungal diversity was not reported to influence the organic matter content of the humus at the end of the experiment.

In another study at the older moist acidic and newer non-acidic tudra regions at the northern foothills of Brookes range, Alsaka by Hobbie and Gough (2004), the higher soil N availability and lower pH at the moist acidic tundra contributed to promotion/greater abundance of soil fungi that resulted in faster decomposition in acidic tundra. The decomposition rates also affected the type of plant species or on ground plant diversity and richness at the two regions. Thus, this suggested that the landscape age and associated differences in soil pH and nutrient availability are important sources of variation in decomposition rate in upland Alaskan tundra.

New tools and techniques for monitoring microbial community succession during composting: New techniques have now been employed to observe the change in microbial forms or microbial succession during different stages of composting process which include use of Denaturing gradient gel electrophoresis (DGGE), T-RFLP, PCR-RFLP etc. The reports dealing with elucidation and monitoring of microbial succession during compost stabilization process by DNA based methods owe an advantage over traditional methods. Michel *et al* (2002) carried out a wide analysis of terminal-restriction fragment length polymorphism (T-RFLP) patterns by comparing the sizes of the fragments from

T-RFLP to those predicted by computer-simulated digestions of bacterial DNA. Similarly Nakasaki *et al* (2005) have reported the use PCR-RFLP for monitoring of microbial community succession during composting process. Ishii *et al* (2000) have reported the analysis of microbial succession during a laboratory-scale composting process of garbage by DGGE combined with measurement of physicochemical parameters such as temperature, pH, organic acids, total dissolved organic carbon and water-soluble humic substance. They reported change in DGGE banding patterns consequent to four phases during composting. The microbial succession determined by DGGE exhibited presence of some fermenting bacteria, such as *lactobacillus* in the mesophilic (S) phase, appearance of thermophilic *Bacillus* in the thermophilic (T) phase and increased complexity as well as diversity of bacterial forms after cooling (C) phase.

References

Hobbie SE and Gough L (2004) Litter decomposition in moist acidic and non-acidic tundra with different glacial histories. *Oecologia*, 140:113-124.

Ishii K, Fukui M and Takii S (2000) Microbial succession during a composting process as evaluated by denaturing gradient gel electrophoresis analysis. *J Appl Microbiol*, 89:768-777.

Krull E, Baldock J and *Skjemstad* J (2001) Soil texture effects on decomposition and soil carbon storage. Nee Workshop Proceedings 18-20 April 2001, pp:103-110.

Michel Jr FC, Marsh TJ and Reddy CA (2002) Bocterial Community Structure during yard trimmings composting. In : Microbiology of Compositing, (eds) Insam H, Riddech N and Klammer S, Springer, Verlag, Berlin, Germany.

Nakasaki K, Nag K and Karita S (2005) Microbial succession associated with organic matter decomposition during thermophilic composting of organic waste. *Waste Manage Res*, 23:48-56.

Neil C and Gignoux J (2006) Soil organic matter decomposition driven by microbial growth: A simple model for a complex network of interactions. *Soil Biol Biochem*, 38: 803–811.

Savoie JM, Minivielle N and Chalaux N (1995) Changes in nitrogen availability and effects of ammonia during composting. *Mushroom Sci.* 14: 275-282.

Setala H and McLean MA (2004) Decomposition rate of organic substrates in relation to the species diversity of soil saprophytic fungi. *Oecologia*, 139:98-107.

Sombroek WG, Nachtergaele FO and Hebel A (1993) Amounts, dynamics and sequestering of carbon in tropical and subtropical soils. *Ambio*, 22:417-426.

Taiwo LB and Oso BA (2004) Influence of composting techniques on microbial succession, temperature and pH in a composting municipal solid waste. *Afr J Biotechnol*, 3(4):239-243.

Mind Bogglers

1. Does a legume green manure crop provide a "free" source of soil N?
2. What is the role of soil organic matter? How the C:N ratio of the soil could affect the soil physico-chemical properties?
3. What are the benefits of addition of compost to particular site? Why is it termed as soil conditioner or soil amendment?
4. What is the significance of microbial decomposers in nutrient recycling processes?
5. Highly substituted phenolics are hard to degrade. Why? Enlist few genera of microbes involved in aromatic compound degradation.

11

Bioremediation

- Introduction 274
- Bioremediation Technologies Available 278
- Applications of Bioremediation 283
- Future Perspectives 291
- Conclusions 294
- References 295
- Mind Bogglers 297

Introduction

Bioremediation can be defined as elimination, attenuation or transformation of polluting or contaminating substances by the use of biological processes, to minimize the risk to human health and the environment. Bioremediation of contaminated soil and water uses microbes (e.g. bacteria and fungi) to convert pollutants to harmless or more environmentally acceptable products and minimizes the risk to human health in a natural or managed process.

Soil contains following microbial forms expressed as cfu g^{-1}:

- ❒ Bacteria : 10^5 to 10^8
- ❒ Actinomycetes : 10^6 to 10^7
- ❒ Fungi : 10^5 to 10^6

This technique applies/utilizes the ability of naturally occurring microbes to degrade a wide range of organic contaminants and in most cases, it relies on different combinations of indigenous microbes (bioaugmentation/dual or mixed inoculums) to deal with different pollutants as the end product (released by one type) could act as a substrate for the other microbial form and hence the xenobiotic or the organic contaminant is mineralized to microbial cell mass and inorganic minor forms (CO_2 and H_2O) under aerobic conditions while methane, CO_2 and H_2 are released under anaerobic conditions.

However, there is a constraint of this biological scavenging system that it can only stimulate the biodegradation of organic pollutants that are susceptible to microbial attack. As the technology develops, the range of chemicals that can be readily degraded is continually increasing which is the need of the hour as human activities have resulted in accumulation of many recalcitrant and hazardous waste compounds/chemicals.

List of chemicals that can be biologically treated

- ❒ Hydrocarbon compounds
  - petroleum hydrocarbons (e.g. petrol, diesel, heating oil, fuel oil, crude oil and lubricants)

- aromatic hydrocarbons (e.g. benzene, toluene, ethylbenzene and xylene (BTEX)
- phenols
- polyaromatic hydrocarbons (PAHs)
- oxygenated hydrocarbons (e.g. glycols, surfactants and detergents)

❒ Chlorinated solvents

❒ Pesticides (including an array of organohalogen/phosphate compounds used as insecticide, herbicide, nematicide, fungicide etc),

❒ Dyes (azo based compounds and other organic dyes)

❒ Explosives (nitroaromatic compounds as napthelene, phenanthrene, anthracene)

The key players in bioremediation are a range of microorganisms as these tinny critters not only have small cellular structures with ability to easily contact contaminant as well as possess versatile enzymatic machinery that help in using environmental contaminant as food by chopping up critical substitutions and chemical bonds. This inherent ability makes them ideally suited to the task of contaminant destruction.

Bioremediation can be regarded as an extension of the purpose that microorganisms have served in nature for billions of years i.e. the breakdown of complex human, animal, and plant wastes so that life can continue from one generation to the next. Bioremediation has distinct advantage over the physico-chemical remediation methods as it is cost effective and achieve the complete degradation of the organic pollutants without collateral destruction of the site material or its indigenous flora and fauna (Timmis and Pieper, 1999).

The first step in any remediation strategy is identification of a region/land which deserves cleaning up processes and then there comes the need or requirement of biological scavenging treatment of the contaminated land. There are several remediation protocols and each one has its own advantages as well as inherent drawbacks and thus application of appropriate treatment method is prerequisite. Bioremediation strategies performing better at one place/site may not work efficiently at other site/place as the key entities in bioremediation i.e. microflora engaged in the degradation process exhibit variabilities in terms of action/ action spectrum.

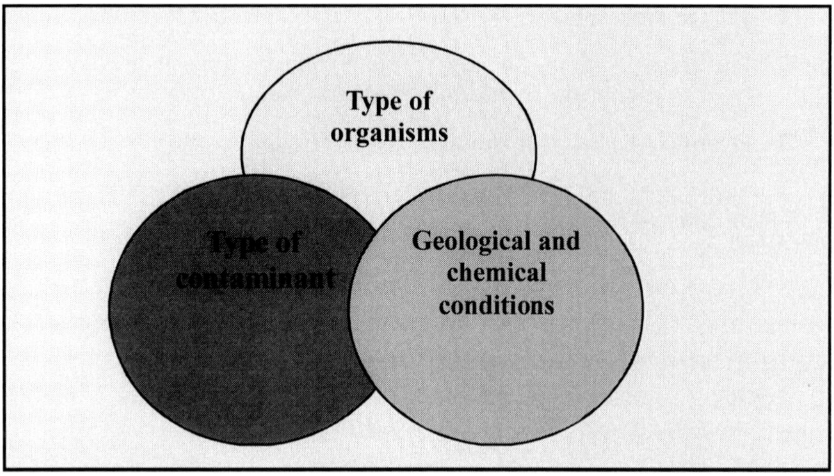

Factors affecting ability of microorganisms to degrade contaminants in the subsurface

Bioremediation strategies

- *Catabolism:* target compound used as carbon source
- *Cometabolism:* target compound enzymatically attacked but not used as carbon source
- *Bioaccumulation:* target compound not metabolized at all but taken up and concentrated within the organism
- *Mineralization:* Complete degradation of organic pollutants (less common process)
- *Partial degradation:* Incomplete degradation of contaminant leading to the generation and possible accumulation of intermediate by-products (general process)

Treatment options can be divided into containment, separation or destruction and these could differ regarding kind of method employed i.e physical, chemical or biological. The flow diagram indicates following remediation protocols that could be employed for cleaning up of different regions while in this section more emphasis would be given particularly on the various bioremediation strategies:-

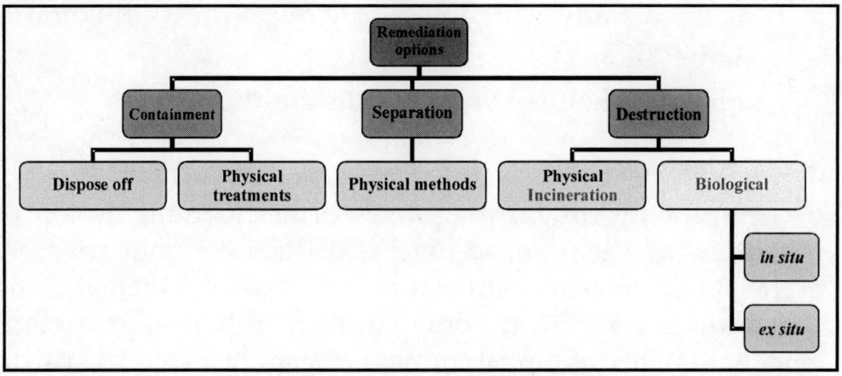

Options for remediation

Bioremediation protocols could also be classified according to the kind of biological entity involved *viz*. plant (Phytoremediation), plant microbial rhizosphere interactions (Rhizoremediation) or fungi (Mycoremediation). The later two have been discussed here while phytoremediation has been discussed earlier.

(a) Rhizoremediation

It could be defined as alternative technique employed for cleaning up of environment contaminated by hazardous pollutants that involves mutalistic relationship between plants and microorganisms (Nichols *et al*, 1997). Thus rhizoremediation depends upon two different classes of living entities *viz* the prokaryotic microbes and the eukaryotic plant cells and thus is liable to variation in effectiveness over a span of time, species of microbe or plant involved, chemical properties/kind of contaminant and climatic conditions. Mackova *et al* (1997) have summarized the role of plants to remediate pollutants for which an array of pathways or mechanisms are followed such as

- direct uptake and conversion of contaminant to non-toxic form followed by its accumulation in the plant cells/tissues
- helper action of plant i.e. release of microbial growth stimulating exudates and enzymes so that an acceleration of the already workable biochemical transformation is followed
- mineralization enhancement in the rhizosphere

(b) Mycoremediation

Employing fungal microbial forms for remediation of contaminated site is called mycoremediation. Fungi are more proficient at cometabolism and bioaccumulation than at using xenobiotics as sole carbon sources and the proficient biodegradability of fungal forms is due to the mycelial growth habit that provides a competitive advantage over bacteria and yeasts. Mycelia rapidly ramify through substrates and digest the substrate due to action of extracellular enzymes which also equip fungal cells to tolerate higher concentrations of toxic chemicals. Few examples of filamentous fungi, *Phanerochaetec hrysosporium* has emerged as the model system for studying xenobiotic degradation. The fungal degradative mechanisms are less explored in general however few enzymes (particularly oxidative enzymes) have been observed to play key roles as well as organic acids and chelators excreted by fungus (lower the pH of the niche surrounding hyphae) help in remediation. Thus fungal remediation has equipped containment bioremediation strategy opportunities for new research and includes the application of various brown rots, litter rots, aquatic fungi, anaerobic fungi and mycorrhizal fungi.

BIOREMEDIATION TECHNOLOGIES AVAILABLE

Bioremediation technologies have been available for treatment of predominantly biodegradable organic contaminants (an array of hydrocarbons, chlorinated solvents, timber treatments, pesticides and explosives) on surface, subsurface and in groundwater. The bioremediation technologies differ according to the type of organic contaminant, amount of area under consideration and type of area (surface, sub-surface, water, sediment etc). Bioremediation of surface and groundwater contamination of organic compounds could be performed commercially using two basic treatment methods *viz ex-situ* and *in-situ* which differ only regarding excavation of contaminated land (soil) the case for the former technique only.

Ex-situ bioremediation technologies: The *ex-situ* technologies are cost intensive owing to excavation and screening procedures required prior to the actual bioremediation step. The technologies exhibit their dependence on soil properties, nature and concentration of contaminants, climatic conditions and the available space for on-site treatment regarding better performance.

This concept includes excavation of the contaminated land/soil/region top soil followed by proper spreading of the contaminated soil onto impermeable surface lined with clay/ceramic tiles/polyethylene (high density) sheets in small piles, windrows (lines/rows) or as layer in open air or in limited space as chamber/closed room and then bioremediation is carried out under conditions optimum for the working and multiplication of microbes. The optimum conditions are maintained by adjusting/taking care of abiotic factors *viz* temperature, moisture and aeration rates by either tilling, turning over of layer/pile or by injecting air in pile/layer/row and spraying of water while the soil pH could be adjusted to neutral by addition of lime or other compounds. The moisture content of the contaminated soil could also be maintained by addition of straw, wood chips or other organic matter that further improves the texture, water-holding capacity, organic matter content and aeration. The initial multiplication of the microbial consortia could be hastened by addition of nutrients which may be mixed with or sprayed over the contaminated soil.

Closed chamber bioremediation: Bioreactor

- used for treating complex mixtures of compounds under controlled conditions
- includes combination of different treatments in a series of reactor vessels
- may include only aerobic/anaerobic steps or both
- combines various chemical and biological processes for remediation of contaminant (s)

In situ Bioremediation Technologies

In situ bioremediation technologies largely depend on the microbial transformation reactions and skills of microbes to use the contaminant or xenobiotic as sole carbon source or as compound in any other step/process of energy generation. The advantage of *in-situ* technologies is that these do not involve the excavation or removal of soil.

In situ bioremediation could occur by increasing the survivality and multiplication of the native vanguard microbial population actively engaged in degradation of the contaminant by addition of nutrients, pumping of air, increasing moisture content, inclusion of plant hosts, inoculation of assisting microflora as sewage treatment slurry, compost etc i.e. biostimulation (Pankrantz, 2001). Bioaugmentation forms an important limb of *in situ* bioremediation technologies, commonly employed and involves approaches for increasing the persistence and activity of exogenous microorganisms and/or genes following introduction into the environment such as use of carrier (usually alginate) encapsulated cells, use of plant host (rhizosphere work as niche for growth and multiplication of introduced microbe i.e. rhizosphere bioaugmentation), engineering gene in carrier inoculant for its movement in microbes already existing in rhizosphere/bulk soil (gene bioaugmentation) or engineering genes in plant used for remediation (phytoaugmentation). Thus *in situ* technologies also include the application of genetically engineered microorganisms (GEM's) that speedup the process of bioremediation.

Bioaugmentation

Process involving introduction of microorganisms able to degrade contaminant/xenobiotic (either derived from contaminated soil i.e. vanguard population or from known stock/preserved culture) into a contaminated system

Biostimulation

Process of introduction of additional nutrients in the form of organic and/or inorganic fertilizers into a contaminated system, which increases the population of the indigenous microorganisms

The *in situ* technologies also involve use of biosurfactants. Several advantages of biosurfacants including biodegradability, specificity, effectiveness, low toxicity (generally), biocompatibility and digestibility, availability of raw materials for biosurfactant production, acceptable production economics and use in environmental control make these compounds a useful tool in major bioremediation methodologies (Kosaric, 2001).

Biosurfactants

- high molecular-weight lipid complexes
- normally produced under aerobic conditions in aerated bioreactors
- few examples are glycolipids, hydroxylated and cross-linked fatty acids (mycolic acids), polysaccharide-lipid complexes, lipoprotein-lipopeptides, phospholipids and complete cell surface.

Thus as a whole the *in situ* bioremediation technologies could be classified as natural (but to an extent monitored) remediation and engineered remediation technologies. The former potentially less expensive technique involves monitoring of natural physical, chemical and biological processes in soil and groundwater to either enhance degradation of contaminant or limit its spread or migration while the later technique includes addition of nutrients and/or air to encourage growth and multiplication of microbes and thus enhanced degradation of the contaminant. The addition of air i.e. bioventing increase

Technique	Advantage	Disadvantage
Ex-situ	Quicker than *in-situ* techniques Useful for shallow decontamination Easier to control Predictable timescales Readily contained in a controlled area Can deal with high concentrations of contamination Remediation conditions can be optimized	Cost intensive Causes enormous disturbance in terms of dislocation of people/ buildings from site of remediation Cannot help in remediation of deep subsurface soil layers
In-situ	Able to deal with deep contamination Avoids expensive excavation Allows remediation around or under buildings without disturbance Reduces worker exposure to volatile compounds Minimizes problems with dust	Tailoring of methods for region/site-specific conditions

degradation as more oxygen enhances the biodegradative capacities of the native microflora however nutrients may also be pumped into the subsurface contaminated layer to further increase the microbial count which is done mostly to remediate unsaturated zone i.e. subsurface layers above the water table. A similar technique of pumping of air via boreholes 'biosparging' could be performed to enhance aerobic degradation of organic contaminants in subsurface layers below water table. Another method injection and recovery system also includes addition of nutrients and air but this is performed in the ground water extracted from the soil. The nutrients and air added water is again injected back to the contaminated site. However the effectiveness of these methods depend on soil permeability, rate of air flow through soil and abiotic conditions optimum for microbial growth (temperature, pH, sufficient nutrients).

Applications of Bioremediation

Last few decades have witnessed rapid advancements in various sectors of human development and among these enormous industrialization resulted in an increase in amount and complexity of different wastes generated and released in the environment. Addition of these wastes/xenobiotics into biosphere has caused a pressure on the soil microflora involved in degradation reactions.

Hydrocarbons and derivatives

Hydrocarbons include the members having hydrated carbon as the skeleton and the prominent hydrocarbons involved in environment contamination and from bioremediation viewpoint are polyaromatic compounds, petroleum and similar compounds. Petroleum (complex mixture of hydrocarbons and other organic compounds) varies widely in composition and physical properties from different reservoirs. These compounds act as carbon source for the microbes and thus could be biotransformed or biodegraded. Hydrocarbon compounds could be utilized by microbes *via* aerobic as well as anaerobic pathways (Lovely, 2000) while several novel catalytic mechanisms for hydrocarbon degradation have been characterized after information big bang in molecular biological/genetical studies that increased our understanding of structural, cellular, biochemical and physiological alterations occurred during hydrocarbon metabolism (Shingler, 2003). A very nice review on recent advances in petroleum microbiology has been written by Van Hamme *et al* (2003). They have reported the characterization of the physiological responses of microorganisms to the presence of hydrocarbons such as cell surface alterations and adaptive mechanisms for uptake and efflux of these substrates. New molecular techniques have helped in investigating the dynamics of microbial communities in hydrocarbon-impacted ecosystems (Alexander, 2000). By establishing conditions which maximize rates and extents of microbial growth, hydrocarbon access, and transformation, highly accelerated and bioreactor-based petroleum waste degradation processes could be implemented.

Bioremediation of hydrocarbon contaminated soil could be initiated by the native microbial community which may be (predicted to be) diverse and complex (as sulfate-reducing bacteria (mesophilic and thermophilic), methanogens, fermentative bacteria (mesophilic and thermophilic) and iron-reducing bacteria in petroleum reservoirs) depending upon vast range of substrates/metabolites (Magot *et al,* 2000).

A wide variety of Polycyclic Aromatic Hydrocarbons (PAHs) are ubiquitous environmental contaminants and are observed to be potent mutagens and carcinogens. These are made up of two or more fused benzene rings and some "pentacyclic moieties" in linear, angular, and/or cluster arrangements and exist in environment as a result of the incomplete combustion of organic matter. However the synthetic PAHs (as naphthalene, phenanthrene) are introduced in environment as these have a role in synthesis of different organic compounds like pesticides, fungicides, dyes and detergents. The hydrophobic and lipophilic nature of these compounds help them to get absorbed in the mammalian intestines and to get accumulated in the lipid bodies or fatty tissues of the organisms.

PAHs are degraded slowly by indigenous microorganisms and persist in soils and sediments. The recalcitrant nature of PAHs is attributed in part to the strong absorption of PAH molecules to soil organic matter and in part to their low solubilities which makes them less amenable to microbial degradation. The major mechanisms involved in PAH utilization and metabolism include the cometabolic transformations and use of these compounds as sole carbon and energy source such as for benzopyrene which acts as a sole carbon and energy source, microbes exhibit cometabolic transformations (Juhasz, and Naidu, 2000, Suenaga *et al,* 2001).

Some organisms capable of degrading Polycyclic Aromatic Hydrocarbons (PAHs)

Polycyclic aromatic hydrocarbons (PAHs)	**No. of rings**	**Organisms**
Napthalene	2	*Alcaligenes denitrificans, Pseudomonas putida, Psudomonas fluorescens, Streptomyces* sp., *Vibrio* sp.
Phenanthrene	3	*Aeromonas* sp. *Alcaligenes denitrificans, Bacillus* sp., *Micrococcus* sp., *Pseudomonas putida, Streptomyces* sp., *Vibrio* sp.
Fluoranthene, pyrene, and chrysene	1-4	*Mycobacterium* sp., *Rhodococcus* sp., *Stenotrophomonas maltophila* VUN 10, 003
benzo[*a*]pyrene	5	*Mycobacterium* sp., *Sphingomonas paucimobilis*

Polyaromatic hydrocarbons

- 16 PAHs listed as priority pollutants by US Environmental Protection Agency (Perez *et al,* 2003)
- Persist in the environment as
  - thermodynamically stable compounds
  - have large negative resonance energies
  - exhibit high sorbtion capacity
  - easily get adsorbed to soil particles
  - produce dead end recalcitrant by-products or intermediate compounds
  - biodegradation interfered by lack of cometabolic or inducer substrate
  - no strains to utilize PAHs with more than four rings

(Juhasz *et al,* 2002)

Bacteria and some green algae initiate the oxidation of PAHs by incorporating both atoms of molecular oxygen, catalyzed by a dioxygenase into the aromatic ring to produce a cis-dihydrodiol, which is then dehydrogenated to give a catechol which is funneled into central pathways of metabolism and could be oxidized to provide energy or utilized for biosynthesis of cell constituents.

PAHs degradation is catalyzed by multicomponent dioxygenase enzyme systems consisiting of

- a flavoprotein (ferredoxin reductase)
- an electron carrier ferredoxin
- a terminal oxidase

Halogenated Aliphatics and Aromatics

Halogenated (chlorinated) organic chemicals/compounds probably constitute half of the environmental organic pollutant problems in the world (Tiejdie *et al,* 1993). Being common pollutants the focus of their bioremediation has been prime question that has cajoled the researchers all over the globe and several reports regarding the degradation of these pollutants have been published. However only recent work has provided requisite information regarding role of individual microbe (whether bacteria, cyanobacteria or fungi) in reduction of the polychlorinated hydrocarbons (PCBs).

Nitroaromatics

Nitroaromatic compounds are recalcitrant contaminants that are mineralized at a relatively slow pace by microorganisms and include nitrobenzenes, *o*- and *p*-nitrotoluene, 2,4-dinitrotoluene, 2,4,6-trinitrotoluene and nitrobenzoates. These chemicals have multiple applications and are involved in the synthesis of more recalcitrant polyurethane foams, herbicides, insecticides, pharmaceuticals and explosives. Microbes could degrade nitroaromatics through several catabolic pathways. Lewis *et al* (2004) have reviewed the whole mechanisms and metabolism of nitroaromatics.

Natural nitroaromatic compounds

- chloramphenicol (antibiotic)
- nitropyoluteorin (purine derivative from the bryozoan *Phidolopora pacifica*)
- oxypyrrolnitrin (antibiotic)
- phidolopin (marine xanthine having antialgal and antibacterial properties)

Trinitrotoulene (TNT)

- more recalcitrant than mono- and dinitrotoulenes
- not attacked or limited attack by classic dioxygenases
- class I flavin-dependent β/α barrel oxidoreductases act as denitrase enzyme

Trinitrotoulene is the most toxic recalcitrant nitroaromatic which is degraded at a very slow pace in the environment. The TNT degradation could occur aerobically as well as anaerobically. The aerobic degradation is carried on by certain strains of *Pseudomonas* and fungi (as *Phanerochaete chrysosporium*) which could utilize TNT as a nitrogen source. The nitrogen in TNT is removed as nitrite by aerobic bacteria followed by reduction of the released nitrite to ammonium while the fungal line of action is via conversion of TNT to reduced intermediates under ligninolytic conditions. These intermediates (nitroso and hydroxylamine intermediates) are excreted to the external milieu to work as substrates of ligninolytic enzymes. Anaerobic microorganisms (*Desulfovibrio* and *Clostridium*) degrade TNT by various pathways but the exact steps in the mechanism are yet to be deciphered however the pathways involve reduction of TNT to either triaminotoluene or hydroxylaminodinitrotoluenes, which are then further metabolized (Esteve-Nunez *et al,* 2001).

Heavy metals and radionuclides

Metal bioremediation using live microbial cells, non-living biomass or biopolymers as biosorbents, is an attractive system regarding lower cost and high efficiency in comparison to physicochemical methods of heavy metal removal. Microbial bioprecipitation and biotransformation reactions (alter several properties of metals *viz.* solubility, sorption characteristics, transport properties and toxicity) on surface as well as in subsurface soil layers help in remediation of toxic metals and radionuclides. The biotransformation of metal ions could be done by several microbial activities however biosorption, bioaccumulation and biologically catalyzed redox reactions leading to immobilization or solubilization are the prime ones. A variety of fungi, algae and bacteria are being actively used as biosorbents for decontamination of heavy metals.

Bioaccumulation involves sequestration, concentration and immobilization of a substance by a microorganism, a process that requires utilization of ATP energy against the concentration gradient while the later is ATP-independent process that involves adsorption of positively charged metal cations to negatively charged molecules in polysaccharides molecules of capsule, slime layer or cell wall surface. The common filamentous fungi can sorb heavy metals (Zn, Cd, Pb, Fe, Ni, Ag, Th, Ra and U) from aqueous solutions to varying extents with biomass regeneration by use of elutants which makes it economical and sustainable option for developing effluent treatment process for removal and recovery of heavy metals. However few recalcitrant metals (Ni) owing to low binding capacity with biomass in presence of organic or inorganic ligands could not be effectively removed from industrial effluents. Moreover metal binding by biomolecules of structural components or secreted polymers is fortuitous and relative efficiencies depend on attributes of the metal ion as well as on the reactivity of the provided ligands (Kotrba *et al,* 1999).

Heavy metal and radionuclide contamination and accumulation of soil results in

- nutrient depletion in the soil
- reduction in availability of nutrients in the soil for plant uptake
- plant uptake of metal ions that mimic some nutrients

- affect solubility of nutrients as some of the contaminants are soluble
- appearance of elemental deficiency diseases in plant
- stunted growth of plant with chlorosis or perforations in leaves and contaminated fruits
- transfer to higher organisms and biomagnification through the food chain

Igwe *et al* (2005)

However solubilization of biosorbed and co-precipitated metals could occur during redox reactions by metal reducing microbes (Gadd, 2004). Microbes exhibit immense versatility in developing metal resistance and thus these microscopic scavengers could be employed for concentration and remediation of the metal contaminated land (Spain, 2003).

Phytoremediation is beginning to attract attention of researchers as plant detoxify heavy metals (cadmium, arsenate and mercury) after uptake which are sequestered in vacuoles (as complexes of metal-phytochelatins) of stem and leave cells. The above ground plant tissue could then be harvested and incinerated to concentrate the heavy metals making it a reasonably cost effective remediation technology. Inoculation of such plants with microbes involved in any of the above activities could help in further increasing the concentration of absorbed metal in the plant tissues.

Heavy metals could also be bioremediated by immobilization of heavy metals into biomass or precipitation by employing microorganism, animals and plants as biosorbents. Specific biosorbents could be designed of high affinity and selectivity for various metals and even the biosorption efficiency could be enhanced several fold by using engineered microbes (may contain more surface receptors or enhanced sequestering enzymes) (Chen *et al,* 2005).

Plastics

Plastics are the organic or polymeric condensation forms (containing carbon backbone adorned with oxygen, nitrogen, sulphur, chlorine or silicone atoms) that include a range of synthetic to semi-synthetic polymerization products. This polymerization reaction decreases the degradation of these compounds in nature and above all the plastic waste in world is increasing at an alarming pace. This demands the remediation protocols for the biodegradation of plastics. The polyethylene form of plastics is most used one and this plastic could be either photo-degraded or chemically degraded. The biodegradation of plastics would require addition of starch, prooxidant or similar compounds that weaken the bonds during polymerization process and these become the site of attack by few bacterial (as *Pseudomonas, Streptomyces* sp) and fungal forms (*Aspegillus flavus, Mucor rouxii*) (Zheng *et al*, 2005). Polyesters and polyolefins could also be degraded by few fungi and bacteria through two steps of depolymerization followed by enzymatic hydrolysis (Howard, 2003, Muller *et al*, 2001). However there occurs a big difference in degradation of aromatic and aliphatic plastic forms.

Recently the plastics of bacterial origin are being synthesized by several researchers. These form the range of bioplastics made up of polyhydroxyalkonates or poly β-hydroxybutyrate compounds (energy storing compounds in bacterial metachromatic granules). These bioplastics could be synthesized on providing either some low cost commercial carbon source or the pretreated or altered plastic e.g. polystyrene utilization by *Pseudomonas putida* for production of biodegradable plastic polyhydroxyalkanoates.

Pesticides

Pesticides include a variety of chemicals such as organophosphates, organochlorines and polyhalogenated compounds. Among these chemicals the organophosphate pesticides (as malathion, parathion) have diverse negative impacts on soil biota (macro and microflora) as well as the higher organisms. These compounds are degraded by aerobic soil microbes. The soil (surface or subsurface) could be excavated from the contaminated site and could be given *ex situ* aerobic bioremediation treatments such as composting and land spreading to enhance degradation of these compounds while groundwater contaminated with organophosphates has to be treated by aerobic bioreactor treatment.

The organochloride pesticides (as chlordane, DDT, toxaphene, gamaxene) exhibit tendency to adsorb on surface of clay/soil particles and also are lipophilic in nature which further increases their chances to get accumulated in the fatty tissues of higher organisms and thus the concentrations get biomagnified at each trophic level in the food chain. These compounds could be remediated by using aerobic followed by anaerobic microbes and thus require a cycling of aerobic (mineralization is completed) and anaerobic phases (crucial as it involves reductive dechlorination these pesticides).

Common fumigants such as 1,3-dichloropropene (1,3-D) and chloropicrin (CP) could be degraded in soil by both chemical and biological mechanisms. Moreover their repeated applications could enhance their biodegradation as a result of selective pressure and proliferance of adapted microorganisms. Several bacteria (prominently physiologically versatile *Pseudomonas* sp) could metabolize/degrade pesticides by using these compounds as sole carbon/energy source however there could also be co-oxidation of these compounds as well during several redox reactions (Dungan and Yates, 2003). The chlorinated compounds as Chloropicrin can be dehalogenated through three successive reductive dehalogenations to nitromethane. Batra *et al* (2006) have reported the utilization of atrazine (common weedicide) as active and sole carbon source by four different bacterial isolates at high (1000ppm) concentration in basal minimal agar medium under lab conditions.

Future Perspectives

Bioremediation process involves a combination of techniques used in many disciplines resulting in development of new techniques over the time along with the enrichment of process techniques and tools of each discipline(s). The acquisition of biodegradative capabilities by native microorganisms at contaminated sites through natural evolutionary processes such as random mutation often occur at an unacceptably slow rates, particularly when multiple biodegradation traits are required (Ang *et al,* 2005). This calls for acceleration of evolutionary processes that could be performed via biomolecular engineering techniques, an increasingly attractive proposition for enhanced bioremediation strategies. Biomolecular engineering involves engineering of the biomolecules (as nucleic acids DNA/RNA and proteins) and biomolecular processes for biotechnological applications which could be done by employing two strategies *viz* rational design and directed evolution (Ryu and Nam, 2000). Among the proficient biodegraders, genes for degradative enzymes of filamentous fungi could be bioengineered in competent fungi or bacteria so that they become equipped to grow in an extended range of environments contaminants (Chen *et al,* 2005).

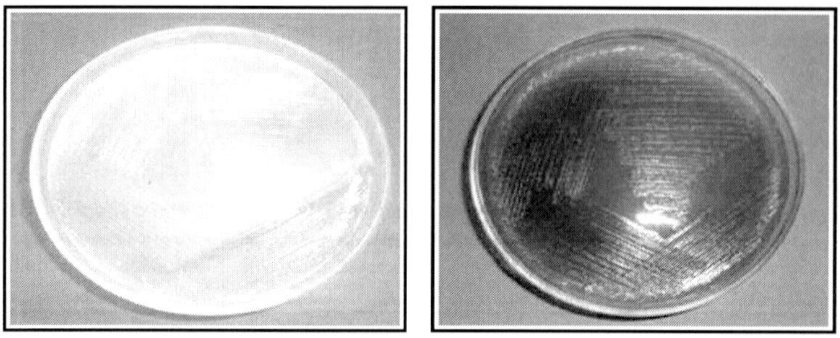

Bacterial isolates on BMN agar supplemented with 1000ppm

New methods of isolation and characterization of novel bioremediating microbes from various sources is becoming a need of hour. Molecular tools would help us in identifying as well isolating new strains/species/genera of microbes that could be utilized as bioremediation inoculants. Leahy and Colwell (1990) discussed molecular tools as colony hybridization and dot blot assays to be revolutionary for describing microbial communities.

Limitations regarding bioremediation

- inadequate understanding of microbial behaviour under field conditions
- difficulty in supply of stimulating materials to microbes
- problems with ensuring adequate contact between microbes and contaminant

Benefits of technological advancement in science and engineering for bioremediation protocols

- Provide better understanding of microbial transformations at ecological, biochemical and genetic levels
- Development/availability of new strategies for bioremediation
- Innovative engineering techniques to for supplying of stimulating substrates/materials to microbes
- Techniques to promote transport of contaminant and enhancement of surface contact of the contaminant with microbe
- Development of GEMs specifically designed for clean up of particular type/class of contaminant

Another latest trend in bioremediation technologies is the development and use of biosensor technology. A biosensor is a device that detects, records, and transmits information regarding a physiological change or the presence of various chemical or biological materials in the environment. Thus it could be used as a probe that integrates a biological component (whole bacterium or a biological product like an enzyme or antibody) with an electronic component to yield a measurable signal (Swensen, 1992). These could be used to monitor changes in environmental conditions in terms of detection/ measurement of altered concentrations of certain hazardous chemicals (pesticides, PAHs, nitroaromatics or other xenobiotics) or pathogens.

Functional parts of biosensors

Three fuctional parts

- biological element (could be whole cell/tissue, bioproduct as enzyme, nucleic acid, antibody, cell receptor etc or biologically derived material) - sensitive in nature
- transducer (transmits signals between biological element and detector element)
- a physiochemical detector element (could be optical, electrochemical, magnetic, piezoelectric probe)

The bioremediation potential of a particular microbial species could be enhanced as well as the span of or range of xenobiotics degraded could be increased by engineering genes from distant species, genera or life forms in one microbe (Pieper and Reineke, 2000). These engineered microbes (termed GEMs) could be utilized for mineralization of contaminant. Biosensors are also useful in well defining or monitoring the fate and transport processes of the organisms (particularly genetically engineered and non-native bacteria) during *in situ* bioremediation of toxicants in the environment which would help in determining the dispersal and effects of these organisms outside the spatial and temporal scales of their intended activity.

CONCLUSIONS

Rampant urbanization and rapid industrialization have resulted in generation of several million tones of toxic waste material. This waste consists of several known as well as xenobiotics which could (themselves or their intermediates) be recalcitrant in environment posing serious health and environmental hazards. Cleaning up strategies for these contaminants require complete mineralization so as to replenish the elements in environment. The physical and chemical methods of remediation of these compounds are very costly and thus bioremediation holds up the flag of cost effective and more environmentally friendly technique. Bioremediation uses natural as well as recombinant microorganisms to break down toxic and hazardous substances by aerobic and anaerobic means. They can be applied on site (*in situ*) or off site (*ex situ*), mediated by mixed microbial consortia and/or pure microbial strains and plants (phytoremediation) or even natural attenuation (Brar *et al,* 2006). Several novel advancements in bioremediation strategies have now equipped us to utilize the benefits of this remediation technology but the maximum use is yet to be harnessed.

References

Alexander M (2000) Aging, bioavailability, and overestimation of risk from environmental pollutants. *Environ Sci Technol,*34:4259-4265

Ang EL, Zhao H and Obbard JP (2005) Recent advances in the bioremediation of persistent organic pollutants via biomolecular engineering. *Enzyme Microbial Technol,* 37:487-496.

Batra M, Gupta RP, Kalia A, Gosal SK and Gangwar M (2006) Indigenous soil bacteria as active degraders of atrazine. Poster presentation at National conference on Frontiers in Biofertilizers and Biopesticides, BAMU subcenter Osmanabad, 19-21 Feb, 2006, Maharashtra, India.

Brar SK, Verma M, Surampalli RY, Misra K, Tyagi RD, Meunier N and Blais JF (2006) Bioremediation of hazardous wastes-A review. *Pract Periodical Haz Toxic and Radioactive Waste Mgmt,* 10(2):59-72.

Chen W, Mulchandani A and Deshusses MA (2005) Environmental Biotechnology: challenges and opportunities for chemical engineers. *AIChE J,* 51(3):690-695.

Dungan RS and Yates SR (2003) Degradation of fumigant pesticides:1,3-dichloropropene, methyl isothiocyanate, chloropicrin and methyl bromide. *Vadose Zone J,* 2:279-286.

Esteve-Nunez A, Caballero A and Ramos JL (2001) Biological degradation of 2,4,6-Trinitrotoluene. *Microbiol Molecular Biol Revs,* 66(3):335-352.

Gadd GM (2004) Microbial influence on metal mobility and application for bioremediation. *Geoderma,* 122:109-119.

Howard GT (2003) Biodegradation of polyurethane: a review. *Internalt Biodeterioration Biodegradation,* 43:7-12.

Igwe JC, Nnorom IC and Gbaruko BC (2005) Kinetics of radionuclides and heavy metals behaviour in soils: Implications for plant growth. *African J Biotechnol,* 4(13):1541-1547.

Juhasz AL and Naidu R (2000) Bioremediation of high molecular weight polycyclic aromatic hydrocarbons: A review of the microbial degradation of benzo[a]pyrene. *Int Biodeterior Biodegrad,* 45:57-88

Juhasz AL, Stanley GA and Britz ML (2002) Metabolite repression inhibits degradation of benzo[*a*]pyrene and dibenz[*a*, h]anthracene by *Stenotrophomonas maltophila* VUN 10, 003. *J Ind Microbiol Biotechnol,* 28:88-96.

Kosaric N (2001) Biosurfactants and their application for soil bioremediation. *Food Technol Biotechnol,* 39(4):295-304.

Kotrba P, Doleckova L, Lorenzeno VD and Ruml T (1999) Enhanced bioaccumulation of heavy metal ions by bacterial cells due to surface display of short metal binding peptides. *Appl Environ Microbiol,* 65(3):1092-1098.

Leahy JG and Colwell RR (1990) Microbial Degradation of hydrocarbons in the environment. *Microbial Reviews,* 53(3):305-315.

Lewis TA, Newcombe DA and Crawford RL (2004) Bioremediation of soils contaminated with explosives. *J Environ Management,* 70:291-307.

Lovely DR (2003) Cleaning up with genomics: Applying molecular biology to bioremediation. *Nature Reviews,* 1:35-44.

Mackova MM, Ocenaskova T, Burkhard J, Demnerova J and Pazlarova J (1997) Biodegradation of polychlorinated biphenyls by plant cells. *J Intern Biodeter Biodegrad,* 39:317-325.

Magot M, Ollivier B and Patel BKC (2000) Microbiology of petroleum reservoirs. *Antonie van Leeuwenhoek,* 77:103-116

Muller RJ, Kleeberg I and Deckwer WD (2001) Biodegradation of polyesters containing aromatic constitutents. *J Biotechnol,* 86: 87-95.

Nichols TD, Wolf DC, Rogers HB, Beyrouty CA and Reynolds CM (1997) Rhizosphere microbial populations in contaminated soils. *J Water Air Soil Pollution,* 95:165-178.

Pankrantz TM (2001) Environmental Engineering Dictionary and Directory. CRC Press, Boca Raton, FL, USA.

Perez S, Reifferscheid G, Eichhorn P and Barcelo D (2003) Assessment of the mutagenic potency of sewage sludges contaminated with polycyclic aromatic hydrocarbons by an ames fluctuation assay. *Environ Toxicol Chem*, 22(11):2576-2584.

Pieper DH and Reineke W (2000) Engineering bacteria for bioremediation. *Curr Opin Biotechnol*,11:262-270.

Ryu DDY and Nam DH (2000) Recent progress in biomolecular engineering. *Biotechnol Prog*, 16:2-16.

Shingler V (2003) Integrated regulation in response to aromatic compounds: from signal sensing to attractive behaviour. *Environ Microbiol*, cited online doi:10.1046/j.1462-2920.2003.00472.x

Spain A (2003) Implications of microbial heavy metal tolerance in the environment. *Reviews Undergraduate Research*, 2:1-6.

Suenaga H, Goto M and Furukawa K (2001) Emergence of multifunctional oxygenase activities by random priming recombination. *J Biol Chem*, 276:22500-22506.

Swenson FJ (1992) Development and evaluation of optical sensors for the detection of bacteria. *Sensors and Actuators*, B11:315-321.

Tiedje JM, Quensen JF, Sanford JC, Schimel JP and Boyd SA (1993) Microbial reductive dechlorination of PCBs. *J Biodegrad*, 4: 231-240.

Timmis KN and Pieper DH (1999) Bacteria designed for bioremediation. *Trends Biotechnol*, 17:201-204.

Van Hamme JD, Singh A and Ward OP (2003) Recent advances in petroleum microbiology. *Microbiol Molecular Biol Revs*, 67(4): 503-549.

Zheng Y, Yanful EK and Bassi AS (2005) A review of plastic waste biodegradation. *Crit Revs Biotechnol*, 25(4):243-250.

Mind Bogglers

1. Differentiate between biodegradation and bioremediation. How *in situ* and *ex-situ* remediation systems differ from each other?

2. What are the benefits of bioremediation strategies?
3. How would you carryout the bioremediation of a naphthalene contaminated site?
4. Heavy metals are toxic in what ways? How would you proceed for bioremediation of heavy metal contaminated site?
5. What are the novel molecular technique(s) for contaminant sensing and upgraded bioremediation?

What are biosensors? What is the principle of biosensors? Which type of biosensors would be most appropriate for remediation of subsurfacel xenobiotics and why?